WITHDRAWN
W9-DJE-127

PROBOTS AND PEOPLE

PROBOTS AND PEOPLE

The Age of the Personal Robot

Timothy O. Knight

McGraw-Hill Book Company

New York St. Louis San Francisco Auckland Bogotá
Guatemala Hamburg Johannesburg Lisbon London
Madrid Mexico Montreal New Delhi Panama Paris
San Juan São Paulo Singapore Sydney Tokyo Toronto

The authors of the programs provided with this book have carefully reviewed them to ensure their performance in accordance with the specifications described in the book. Neither the authors nor McGraw-Hill, however, make any warranties whatever concerning the programs. They assume no responsibility or liability of any kind for errors in the programs or for the consequences of any such errors.

PROBOTS AND PEOPLE: The Age of the Personal Robot

A BYTE Book.

1 2 3 4 5 6 7 8 9 10 SEMSEM 8 9 3 2 1 0 9 8 7 6 5 4

ISBN 0-07-035106-6

Library of Congress Cataloging in Publication Data

Knight, Timothy O.
Probots and people.

Bibliography: p.
Includes index.
1. Robotics. I. Title.
TJ211.K59 1984 629.8'92 84-5794
ISBN 0-07-035106-6
The sponsoring editor for this book was Jeffrey McCartney; the editing supervisor was Margery Luhrs.

To Yen-Chi, the most brilliant
and beautiful girl in my world,
and the one who has kept me
from becoming a robot myself.

CONTENTS

ACKNOWLEDGMENTS x

INTRODUCTION xi

CHAPTER ONE
THE AGE OF THE PROBOT 1

The Heart of a Probot 3
Are We Ready for Probots? 6
Humans and Robots 10

CHAPTER TWO
FROM FACTORY TO FAMILY 13

What Is an Industrial Robot? 14
The Importance of the Industrial Robot 16
The Advantages of Industrial Robots 17
The Dark Side of Robots 19
The Three Stages 20
The Creation of Jobs 22
The Human Factor 25

CHAPTER THREE
HEROES ALWAYS WIN 27

Basic Facts about HERO 1 30
Operating HERO 1 31

Making HERO 1 Talk 43
The Potential of HERO 1 46

CHAPTER FOUR
MY THREE PROBOTS 49

The Android Robot 50
AndroMan 52
F.R.E.D. 52
Topo 55
Programming Topo with TopoBASIC 57
Calibrating Topo 61
The Useful Probot 64
Topo and the Future 65

CHAPTER FIVE
B.O.B., RB5X, AND MAKING PROBOTS 67

B.O.B., Brains or Brawn? 68
Uses for Brains On Board 71
The RB5X Probot 72
Meet the RB5X 72
The Future of RB5X 80
Robotic Research 81
The Probots in Review 88

CHAPTER SIX
THE PRACTICAL PROBOT 91

Utilizing the Probot's Power 92
Uses for a Probot 95
The Potential for Programmers 100
Imagination and Probots 101

CHAPTER SEVEN
PROBOTS FOR TOMORROW 103

The Future of the Robot 104
New Industries 106
The Effects of Probots and Robots 107

Negative Side Effects 111
The Bright Future 112
Probots, Today and Tomorrow 113

BIBLIOGRAPHY 115

GLOSSARY 117

INDEX 121

ACKNOWLEDGMENTS

I would like to give my appreciation to these companies for their assistance and encouragement during the creation of this book: Heath Company, Androbot, Inc., Nina Stern Public Relations, and RB Robot Corp. Special thanks go to Joel Schwartz, Eastern Regional Manager at Androbot and to Andy Wells, consultant and reviewer for McGraw-Hill for their developmental and promotional assistance.

INTRODUCTION

This book is about today and tomorrow, work and leisure, production and consumption. It is also about a technological revolution that will bring exciting new changes to our lives beyond those of the computer and will create greater prosperity than we have ever known. This book is about the personal robot, or the *probot*

The probot is a recent introduction in the world of consumer electronics. A few years from now, the personal robot will actually be used in millions of homes in the United States and abroad, just like the home computer is today. However, at the present time, like the home computer of 1977, the probot is thought by many to be little more than a sophisticated plaything, capable only of amusing guests at parties and attracting attention at shopping malls.

It is not difficult to imagine why so many people are skeptical. In their view, the probot is at most a computer with wheels attached to it which will steal jobs from hard-working laborers. These same skeptics probably thought that the personal computer would destroy the minds of those who used them. As we have learned, however, computers can actually improve the mind; they are nothing more than machines which make life easier, more productive, and more enjoyable.

Personal robots, or any robots for that matter, were invented for these same purposes. As you read this book, you will discover that robots increase productivity, lower prices, improve prosperity, provide more leisure time, and actually *create* jobs. Even today, the personal robots available can perform a multitude of tasks, from guarding your house at night to teaching you more about computers.

Probots are exciting, expandable, fascinating machines which will grow in number, drop in price, and increase in sophistication over the next few years. This book is intended to introduce you to robotics, to help you become familiar with the idea of personal robots, and to describe some of the probots currently available. It also suggests some of the effects probots could have on our future, along with some ideas about the use and gradual improvement of the probot as a widely-accepted, essential household appliance.

Although most new ideas seem strange at first, I hope that, after reading this book, the idea of a probot in your own home will seem exciting and meaningful. Robotics is going to play a big part in all of our lives, and it's a good idea to be ready for this latest technological revolution. This book is your introduction to robots, both personal and industrial, and to the bright future they will bring us.

PROBOTS AND PEOPLE

CHAPTER ONE
THE AGE OF THE PROBOT

Imagine a morning, not many years from now, on which you awaken after a long and safe slumber. The night before, you drifted to sleep with absolutely no fear that your house would be robbed or that you would be in any danger. Your household computer informs you that your breakfast is ready, so you proceed to the kitchen to enjoy the meal that your kitchen robot has prepared.

As you eat, you glance at the headlines of the morning news on the nearby video monitor, noting that unemployment has been virtually eliminated, the gross national product has increased 15 percent over the past year, and your stocks are doing well. Before you leave, you remind the household robot to test the children on their math lessons, review them on their other studies, and finally, join them in a game of catch. You then step into your transportation vehicle, which whisks you away to the office for a full four hours of work. As you plan the remainder of the day, you decide to spend the rest of the afternoon playing racquetball with some friends, followed by some good reading. After you eat your already-prepared dinner with your family, you will work with one of the household computers for the rest of the evening, then drift off to a pleasant sleep once more.

A day like this would make many people think that they were in Utopia, yet this little excerpt from the life of one man is not as impossible as it might sound. In fact, parts of the preceding story are already true for some people. The factors guiding us into this more productive, less strenuous, and much more leisurely life-style are all centered around one thing: new technologies. These exciting advances, just emerging from the laboratories and factories, include bioengineering, personal computers, and extensive space exploration. However, a personal robot is one machine that will probably have a more profound and widespread effect on society than any other invention. The dream of this creation has been portrayed in plays, movies, and television, but only now is this exciting new tool beginning to demonstrate its potential.

Of course, news stories about the giant industrial robots in factories are seen frequently, and although these machines are efficient, productive, and useful, our lives have not yet been altered dramatically by them. In fact, there are relatively few robots actually working in factories. If this is the case, how will robots make such a dramatic effect on so many people?

The fact that they *will* have a major effect on industry will be discussed later. But the type of robot that will cause the greatest changes is not an industrial, but rather a personal robot. Or, as I like to call it, a *probot.*

Probots will have a more immediate and direct effect on us than industrial robots because personal robots will actually be in the home, doing household chores, teaching the children, and even walking the dog. Although industrial robots will eventually have a profound effect on both labor and productivity, it simply takes a longer time for their full impact to be felt.

THE HEART OF A PROBOT

The robot and the computer are very closely related. In fact, the robot may be thought of as an extension of the computer. In the same manner, the probot may be thought of as an extension of the personal computer. That is, a personal robot is a personal computer, and a great deal more.

Like a home computer, a probot does not have human intelligence, but it can be programmed to perform many tasks, such as teaching lessons, remembering and organizing large amounts of information, and entertaining people.

However, because millions of people already own home computers, the abilities of these machines are familiar to almost everyone. The power of the probot is still a mystery to many people, but its abilities (like any kind of robot) are really not difficult to understand.

A probot is basically a personal computer with a body (Figure 1-1), and if equipped with wheels, is enabled to move from one place to another. And, if the body has an arm, the probot may pick up, move, and manipulate objects. The power of the probot to move itself and to use an arm gives it tremendous potential.

Probots can be made with so many other features that, before long, the home computer begins to seem like nothing more than a handicapped robot. To help you understand what a probot (and a robot) can do, here are some of its most basic components, along with the uses for each feature:

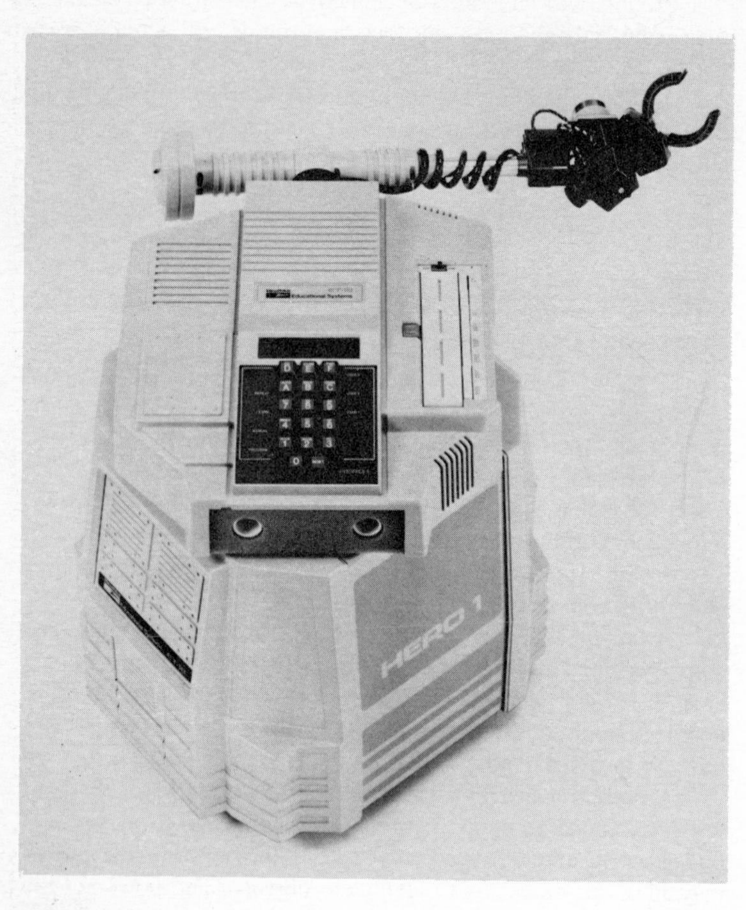

Figure 1-1 A probot is basically a computer with a body. *(Courtesy of Heath Co.)*

Mobility: The probot can move around, being careful not to bump into walls or people along the way. If it does run into something, however, it will remember the mistake and avoid the obstacle in the future. Because the probot can come to you, there is no need for you to go to or even search for the probot.

Arms/Grippers: Because it has the ability to pick up objects, the probot can change the positions of things, carry them from one place to another, or clean up a room. Just think how many tasks your arms perform and it will give you an idea of the additional capabilities given to a probot.

Voice: Many probots (and robots) can be programmed to speak, which allows them to give you information for many purposes: teaching you a foreign language, informing you that its battery power is running low, or alerting you that a burglar has entered the house. This verbal communication is a very important tool.

Sensors: The ability to detect light, heat, sound, infrared rays, and distance gives the probot many new uses. With these sensors, the probot has "eyes," "ears," and "sensation," along with the power to perceive the distance between itself and an object. It could warn a person about a fire with its heat sensor, or it could detect a robber with its sound sensor. This computer on wheels could even greet visitors in your home by measuring the distance between itself and the humans, then rolling over to say "Hello" with its voice.

Programmability: Because probots (and robots) usually have a large amount of memory, they can be programmed for a great variety of tasks. For instance, a probot could be programmed to teach a child about math, using its voice to ask the problems and allowing the child to punch in the correct answer. Probots can also store a great deal of information within their memories, so they can recall the layout of a house, storing the paths of movement from room to room which have been programmed, or they can remind you of an important meeting you have at nine-thirty in the morning.

With these basic features, and the extras available for some probots, the uses for these amazing machines are numerous. Some of the uses have been named here, but a few others include having the probot carry beverages to guests, read stories to a child, serve as a babysitter, play games, or even teach people the fundamentals of computers and robotics. In fact, the probot is such an interesting machine that it could be a very effective advertisement for a store, reciting a preprogrammed speech while moving about in a crowd of people. And just think of how marvelous it would be to have a servant without having any guilt; a probot is the answer.

ARE WE READY FOR PROBOTS?

It seems that with every new invention, discovery, or technology, comes a group of skeptics ready to shoot down anything different and exciting. But not everything stays the same, and the cynics soon realize that many things they thought would surely fail have, in fact, succeeded.

The telephone, for instance, was regarded by many as a temporary plaything of no real usefulness. However, business and general communications today would be impossible, or at least very difficult, without the telephone. In the same way, the television seemed insignificant when it was first emerging, and it was hard to imagine that the new medium would replace radio as our main source of entertainment one day. However, as time went on, television sales boomed, and the skeptics sought something else to doubt. More recently, the home computer was thought of as a toy for the affluent, with no real purpose. Today, millions of people would have far less efficient and productive lives were it not for their computers. This trend will continue, for many things regarded as luxuries eventually become necessities.

This luxury-become-necessity idea will probably hold true for probots as well. The probot, of course, is in its infancy, but its growth, both in numbers and in sophistication, should follow much the same path as that of the home computer. In fact, now that people are becoming more receptive to high technology, the probot revolution might happen even more rapidly than the personal computer revolution did.

In this case, history will probably repeat itself, so to have a look into the future, we should re-examine the events which made up the home computer revolution. In this way, we can imagine the stages of acceptance which probots will most likely go through (Figure 1-2).

HOBBYIST INTRODUCTION
(home computer: 1975; probot: 1981)

This stage was mainly for electronics buffs and other persons interested in technical experimentation. These people formed clubs so that they could trade information for the particular technology in which they were interested.

Today, there are already several groups in the United States who work with probots and share information. Sometimes these hobbyists even make probots themselves and spawn interest in the new, emerging technology. Although this is certainly innovation on a small scale, it does serve as a starting place.

MEDIA BOOM
(home computer: 1976; probot: 1982–83)

Once a certain creation or invention seems interesting enough to capture the public's attention, the media begins their part of this process. By writing stories about the technology—in this case, that of home computers—newspapers, magazines, and television shows made the public aware of personal computing. This coverage made some people so excited about the new invention that they wanted to purchase one. The books and articles predicting the personal computer revolution were believed by some people but shunned by many—though most of the predictions came true eventually.

In the same manner, the media has begun to pay far greater attention to personal robots, which makes the probot revolution fit the pattern of the home computer revolution up to this stage.

MARKET INTRODUCTION
(home computer: 1977; probot: 1983)

Once enough interest was generated by the media, several companies began making home computers. These companies in-

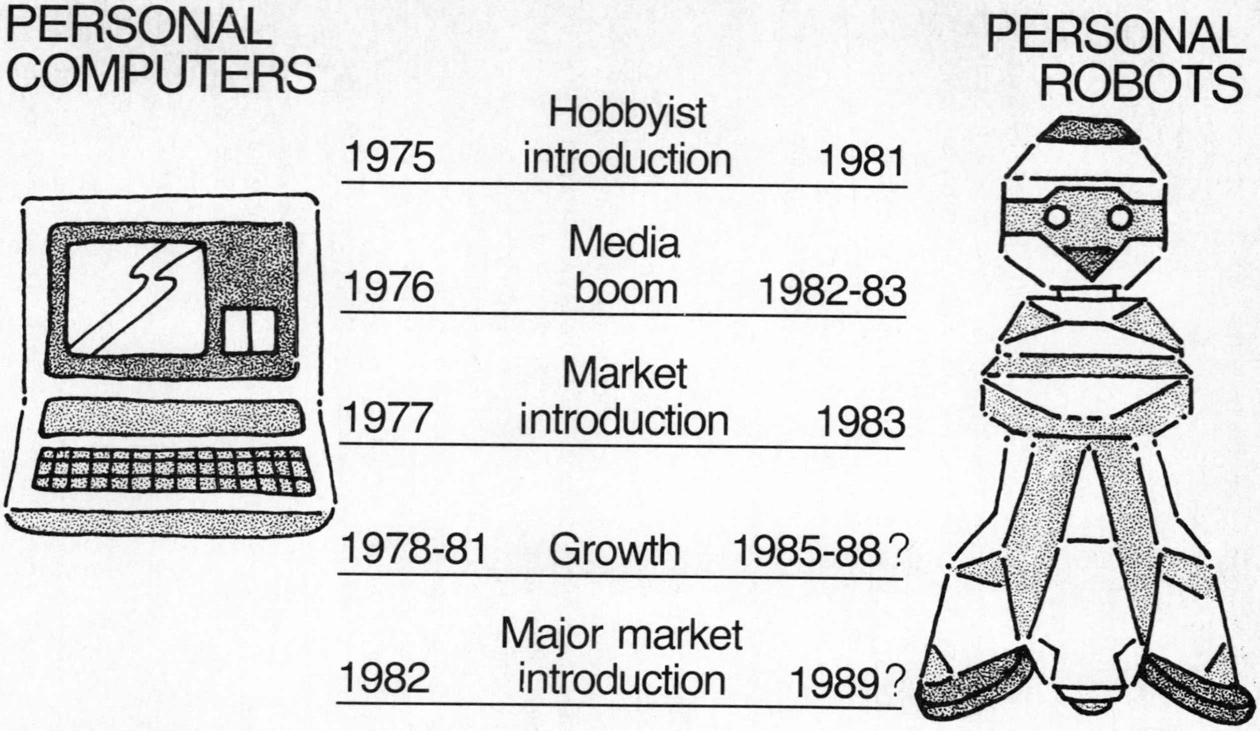

PERSONAL COMPUTERS		PERSONAL ROBOTS
1975	Hobbyist introduction	1981
1976	Media boom	1982-83
1977	Market introduction	1983
1978-81	Growth	1985-88 ?
1982	Major market introduction	1989 ?

Figure 1-2 The personal robot revolution will be similar to the personal computer revolution.

cluded Tandy Corporation, maker of the TRS-80 computer, and Apple Computer, Inc. The computers were good for their time, and made a small dent in the potential market for computers. This initial market introduction sold computers in the hundreds of thousands, but only began to touch the large number of people who could use personal computers.

Just as Apple and Tandy were the first major manufacturers of home computers, Androbot, Inc. and Heath Company are the first major manufacturers of probots.

GROWTH
(home computer: 1978–81; probot: 1985–88?)

The best indication of what will happen to probots next can probably be found in the final two stages of the home computer revolution.

Once a large number of people had purchased computers, companies began seeing the profit potential and began making computers, software, and computer peripherals of their own. Although few of these companies became very large in the computer business, they did serve the purpose of providing healthy competition for the larger companies, and they also created software for Tandy and Apple computers. More people began to get involved with home computers, until approximately a million people owned one of these machines.

MAJOR MARKET INTRODUCTION
(home computer: 1982; probot: 1989?)

In this final stage, the giant companies, such as IBM, saw that the personal computer was worth manufacturing. Also, prices of home computers fell so low that nearly anybody could buy one. Computers such as the VIC-20, which could be purchased for under $100, sold over one million units, while more sophisticated computers, such as the Commodore 64 and Atari 600XL, could be purchased for under $200. At that point, the sales of computers snowballed and now computers are in millions of homes in the United States. As the trend continues, only the very strongest computer manufacturers will survive; companies supporting the computers with software and supplies should prosper. The revolution is over, and the computer is more or less accepted as an important tool within the home.

As you can see, technological revolutions pick up pace very quickly, and since the probot appears to be following the steps of its predecessor, the personal computer, it is likely that stages four and five are going to apply to the probot as well.

These five stages also demonstrate that one of the greatest advantages of the free enterprise system is competition. In the early days, a home computer would cost well over $1,000. However, just a few years later, a far more powerful machine, with a large software library as well, was available for under $100. There is no question that the $2,500 probot will be purchased only by those affluent enough and interested enough to acquire one, but there should be little doubt that in only a few years, an even more powerful, and far less expensive probot will be on the market.

HUMANS AND ROBOTS

Even though the personal robot is not yet widely available, almost everyone has a particular idea about the robot. Some may see it as a threatening creature, whose only purpose is to take away the jobs of helpless humans, or even destroy humans for its own purposes. Others may see robots as cute little machines, somewhat like mechanized Laurel and Hardys. There are also some who imagine robots as efficient, accurate, cost-efficient machines that can increase productivity and will never go on strike.

There are many factors which contribute to our views about what a robot is like. The robot, however, is not as recent an idea as one might expect. The word "robot" was first used in 1920 by Karel Čapek in his play, *R.U.R.* (Rossum's Universal Robots). The term came from the Czech word *robota,* meaning dull labor. In the play, machines called robots are made to work in factories, releasing humans from the "degradation of labor." The maker of the robots points out that they will produce so much that humans will no longer have to worry about working hard to be fed and sheltered. However, as robots spread around the world, governments turn them into soldiers to fight each other. To end this, the robot makers give the androids intelligence and sophistication. Unfortunately, the robots then realize how oppressed they have been, and rebel against the humans. One of the last lines in the play is, "Nobody can hate man more than man."

R.U.R. probably made people feel that robots were dangerous and volatile, and this idea has remained with us for years. Science fiction movies have portrayed the robot as a "heavy," in films such as *Robot Monster* (1953), *Colossus of New York* (1958), and *Silent Running* (1972). Interestingly enough, the movies in which the robots are evil tend to be cheap B-movies, while the high-quality films such as *Forbidden Planet* (1956) and *Star Wars* (1977) portray robots as friendly and human-like.

Unfortunately, many movies, comic books, and novels have made us imagine that robots are to be feared. One book, however, which portrays robots as good rather than evil is Isaac Asimov's *I, Robot,* a collection of his short stories about the machines of the future. Written several decades ago, Asimov's stories seem almost prophetic, since his descriptions of the robotic technology are relatively accurate. Despite this accuracy and the excellence of the stories, the dates are not quite right, because the first robot (a monstrous, voiceless giant named Robbie) is supposedly invented in 1996.

The most memorable aspect of *I, Robot* is probably the three laws of robotics. In the book, these laws are built into every robot, for our safety and for theirs. They are:

First Law: A robot may not injure a human being, or, through inaction, allow a human being to come to harm.

Second Law: A robot must obey the orders given it by human beings, except where such orders would conflict with the First Law.

Third Law: A robot must protect its own existence as long as such protection does not conflict with the First or Second Laws.

The value of these laws is apparent. If such laws were an integral part of every robot manufactured, most people would probably stop worrying about the possible danger robots present.

Our views of robots vary from one person to another, yet the fact that many poeple are somewhat afraid of robots could be a problem. Of course, several years ago, many people were afraid of computers, but now only a few remain who are truly terrified of the machines. Fortunately, specialists have been trained to ease a person's fear about the computer by proving it is a helpful yet harmless tool. In the same manner, many people will probably fear robots at first, but, eventually, society will grow accustomed to them, and realize that robots were made to serve without harming or ruling over humanity.

CHAPTER TWO
FROM FACTORY TO FAMILY

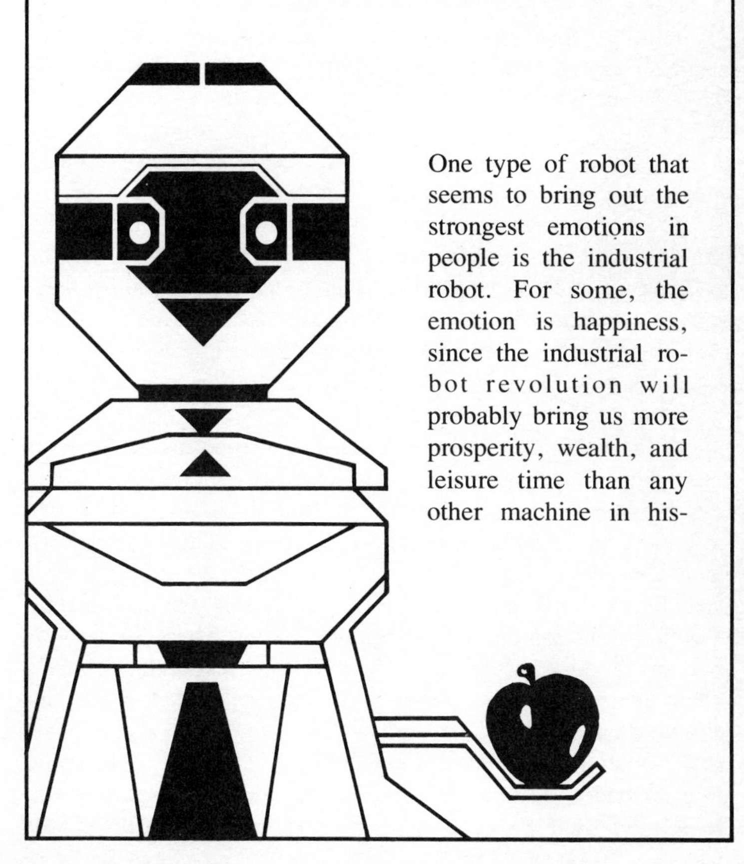

One type of robot that seems to bring out the strongest emotions in people is the industrial robot. For some, the emotion is happiness, since the industrial robot revolution will probably bring us more prosperity, wealth, and leisure time than any other machine in history. For many others, however, the idea of an industrial robot seems frightening, strange, and even dehumanizing.

It should be made clear the industrial robots are not a threat. In fact, these machines will be important to our well-being, since their absence would bring us enormous problems, both social and economic. The fear that robots in the factory will take away jobs is understandable, but really not valid, as I hope this chapter will prove.

WHAT IS AN INDUSTRIAL ROBOT?

An industrial robot is, in its simplest form, a mechanical device which can be programmed to perform a useful industrial task. Even a mechanical arm that moves boxes from one part of a building to another is considered an industrial robot, as long as that machine has been programmed for that task, instead of being constantly controlled by a human.

Currently, robots are usually programmed to do one specific, monotonous task. Programming a robot involves instructing the machine where to move its arm, when to tighten its grippers, and how to twist its arm. By programming the robot down to the most minute movement, just about any specific task can be accomplished any number of times. In fact, on an assembly line, the robot could very well perform a task day and night, every day of the year. This makes the concept of robot assembly lines interesting, since production could be greatly increased.

Because most robots aren't very "intelligent" yet, they can't perform very complicated tasks such as assembling pieces of another machine. Robots are usually programmed to do the most basic tasks, such as spot welding automobile bodies, loading and unloading equipment, and spray painting. These are all jobs that can be done in the same mannner for a particular product, which makes the robot productive, precise, and cost-efficient (Figure 2-1).

Probots and industrial robots are similar in that they can be programmed to do specific tasks, but the probot tends to be more flexible than today's industrial robot. The probot has the power

Figure 2-1 Today, industrial robots are used mostly in the automobile industry.

to talk, move its body as well as its arm, detect light, heat, sound, and serve as a mobile refrigerator. The industrial robot, on the other hand, costs about twenty-five times as much, and usually can only move its arm and hand (gripper). However, the industrial robot can lift many times the weight a probot can, and it is able to move its arm and gripper with great precision. Industrial robots are built for only one specific purpose, but they perform that task with precision, reliability, and consistency, which demands more electronics and labor from its creators.

THE IMPORTANCE OF THE INDUSTRIAL ROBOT

There are quite a number of things that attract manufacturers to the industrial robot. First of all, the robot will never get sick (except perhaps for an occasional breakdown). The robot cannot be harmed by radiation, dust, or contamination. It will never grow weary, which means it can work constantly. Robots don't go on strike or demand wage increases or vacations, and they don't complain to management about boring work or bad plant conditions. More importantly, robots do their work with extraordinary precision, and can be reprogrammed to do a different task when production demands change.

Why, then, doesn't every manufacturer in America equip his factory with an army of robots? Mostly because robots today can only perform menial tasks, and a large part of manufacturing requires judgment and the skilled labor of humans. Manufacturers are waiting for more adept robots. But should they wait too long, they will find themselves trailing far behind other industrialized countries; robots are very much a key to the prosperity of our economy in the future.

The United States doesn't have very many industrial robots—only about six thousand at last count—while seemingly more efficient countries, such as Japan, have over thirty thousand robots at work. One interesting conclusion that can be drawn from these figures is that Japan, which has a very low unemployment rate, also has the greatest number of robots. It seems, then, that robots have not had a negative effect on human employment so far.

Although industrial robots have not reached very popular levels with American manufacturers, it should be only two or three years before the advantages of the robot are fully realized, and the fears that workers sometimes have about them are diminished.

THE ADVANTAGES OF INDUSTRIAL ROBOTS

Besides the fact that the robot is a tireless, cost-efficient, nearly perfect worker for some jobs, there are still other advantages that make the concept of industrial robots even more appealing, both to manufacturers and to the general public.

In many instances, one robot can perform the work of several humans. In fact, in a factory filled with robots, a relatively small number of people is needed to run the plant. One such complex, the Yamazaki Machinery Works in Kentucky, produces the equivalent of a factory employing hundreds of workers, yet only six humans are required to maintain the multitude of robots within the Yamazaki factory.

Robots can also perform some tasks more quickly than humans. Arc welding, for instance, is a dirty, dangerous job which makes it necessary for a worker to wear protective clothing. The job is accompanied by so much smoke and sparks that a worker can only bear to weld 25 percent of the time. A robot, however, which is practically impervious to heat and smoke, can weld 75 percent of the time. In other words, the robot can produce three times as much as the human.

If the ''mindless'' robots of today can accomplish so much, the possibilities for the more sophisticated, intelligent robots of the future seem almost limitless. Intelligent robots would have the ability to hear, see, and feel as much as is possible for a robot. These machines could take the appropriate actions when unforeseen circumstances arise. For example, today's robots require every group of parts they assemble and every piece of equipment they move to be precisely in place, since the robots are programmed for very specific tasks. They are instructed to perform certain actions, such as moving an object from one place to another, based on definite assumptions such as the object will be exactly in place when the robot tries to move it. If it is misplaced or absent altogether, the robot will simply try to perform its task, probably causing such a mistake in production that an entire assembly line could fail.

On the other hand, an intelligent robot would be able to detect problems on an assembly line, and could act accordingly. If an object were out of place, the intelligent robot could simply reposition the object to be properly placed. Or, if the object were missing, the intelligent robot could alert a human operator about the problem.

Intelligent robots have other applications as well. Instead of having to perform monotonous work which a chimpanzee could probably accomplish successfully, the intelligent robot could be programmed to perform more complex tasks. They could put parts of an engine together, supervise "dumb" robots, or be ready to take the place of any robot that breaks down.

Another exciting thing about intelligent robots of the future is that Americans are actually ahead of the Japanese in making these super robots, and it is expected that Americans will have manufactured thousands of them by 1990.

One question that plagues workers when the subject of robots is introduced is, "Are we going to lose our jobs to robots?" The answer to this question will be answered in detail later in this chapter, but it should be said now that a distinct advantage to industrial robots is that productivity problems in the United States (and other countries, for that matter) could be solved more easily with the use of robots. Production can be increased greatly by utilizing robots, and, as we will see, humans will not necessarily be out of work, but perhaps be free to do more creative jobs. More importantly, if the United States does not take an active part in the full-scale use of industrial robots, we will probably fall far behind foreign countries in production and the quality of our products. In that case, unemployment would be an even greater problem than it is now. Therefore, instead of asking if robots are a threat to their jobs, workers should ask instead,

''Why aren't we using robots as much as other industrial countries?'' since robots will be vital to the well-being of productivity and employment in the United States.

THE DARK SIDE OF ROBOTS

Robots are not the solution to all of our problems, though. It would be dishonest to suggest that robots will not bring problems of their own, as all changes do. And robots are going to cause some very big changes in the way we live, work, and play.

Although, as we'll discover later, robots will create many more jobs for humans, there will indeed be jobs lost. Some studies indicate that 25 percent of the industrial work force could be removed from work in the 1980s. Of course, some of these workers will find better jobs, but not all of them will. It is also estimated by the Massachusetts Institute of Technology that some 100,000 workers will be replaced by 32,000 robots in the automobile industry alone.

The idea of workers being replaced by machines is nothing new to the world, though. During the Industrial Revolution, many workers were replaced by new, efficient machines, and this exciting and productive period of history also caused a great deal of poverty, unemployment, and upheavel, just as any major economic change does. Robotics, the second industrial revolution, will probably cause similar changes, though not nearly so dramatic and devastating as those of the Industrial Revolution in the nineteenth century.

The persons who will probably be hit hardest by the industrial robot revolution are the laborers or blue collar workers, who make up 24 percent of the employed. As automation and innovation improve, more of these blue collar workers, especially unskilled laborers, will find themselves either unemployed or in different types of work. Fortunately, since those entering the work force tend to be increasingly educated as years go by, more advanced jobs such as robot maintenance could be filled by people who otherwise would have had the less interesting job of an unskilled worker.

Despite the fact that those people technologically unemployed, or replaced by, industrial robots will probably find other jobs, there will still be a period of transition when unemployment could be a problem. The important thing for the management of companies to remember is that many workers could be moved from the dull jobs which the robots will perform to more exciting, challenging jobs, so a great deal of complaining and strife about the introduction of robots could be diminished. The management of any company that decides to use robots on a large scale must also remember the worker, or else workers will resent the robots, perhaps stalling the industrial robot revolution even more, and causing the United States to fall behind foriegn countries in productivity and technology.

THE THREE STAGES

Industrial robots will go through three main stages as they continue to improve and become more sophisticated. Each of these stages will take roughly a decade each, and will gradually make a nation of workers and management into a nation of people with plenty of free time for leisure activities. These stages are:

1. **Introduction.** At this stage, robots are for the most part put into existing factories and manufacturing centers where they are made to do repetitive tasks. The factory is not designed around the robot, so the robot must be adapted to the workplace. Many robots will work all day and night, seven days a week, with little human supervision. Of course, more complex jobs will still be handled by humans, but as the robots grow more advanced, fewer workers will be needed to make a product.
2. **Modification.** Once robots have been adapted to the factory, the factory begins adapting itself to the robots. This is for the purpose of making the factory more efficient for the robot's sake. A great deal of employment is generated here, since designers, engineers, builders, new management, and maintenance are all required to modify and maintain the modified, more efficient factory.

3. **Self-Reproduction/Robot Explosion.** Once robots reach a certain level of sophistication, they will actually be able to design and build other robots. In fact, the robots they design will be less expensive and more productive, so that costs will decline as fast as productivity increases. This is the most exciting of the stages, since the robots require almost no human assistance, and they are actually self-generating machines. The robots will be able to handle every aspect of manufacturing, and other robots can begin to explore other areas of work, such as mining the seabed for food and minerals, exploring outer space, and constructing giant structures in space, such as solar energy satellites and space stations. The possibilities for robots at this point boggle the mind: it is very difficult to predict what this third stage of the industrial robot revolution will bring us.

Currently, of course, we are only in the first stage, since our robots are relatively unsophisticated, and industrial robots have not had a major impact on the way we work or explore unknown resources. In fact, the robot itself is still an unknown resource, since much of its potential and power are a mystery to the majority of the population.

The rate at which robots are improving is rather slow at this point, largely due to the fact that funding for robotic research is relatively modest. There have been no truly major efforts by government or industry to develop robots, though the funding for robotics will almost certainly increase as the importance of this advanced automation becomes apparent.

Robotics research is currently a long-range time-consuming, very risky venture. It has been estimated that the transformation of our society from human workers to robot workers will cost the entire wealth of this country. In other words, this transformation is going to require all of the capital we presently possess. After the transformation has been made, though, the growth and development of robotics will grow exponentially. Robotics research is indeed a long-term investment, but it promises to be very profitable in the long run.

THE CREATION OF JOBS

Robots do not destroy jobs; they create them. This statement may seem strange to anyone who assumes that since robots do the work of humans, people will be left without jobs. This assumption is incorrect, though, because

1. There is not a fixed number of jobs.
2. The development, manufacture, and maintenance of robots requires human work.
3. After the robot industrial revolution is complete, there will be enough wealth so that humans will not necessarily have to work.

It is understandable that new technology brings fear to workers. Machines always seem to be threatening the jobs of humans, and technological unemployment is growing more frequent as the sophistication of our machines grows. Still, machines haven't taken the place of humans, as the nineteenth-century Industrial Revolution showed. It must be remembered that, despite the invention of machines such as the cotton gin, the reaper, and the computer—each of which took the place of many human workers—the net result was an increase in productivity, not a decrease in employment. Machines may cause the temporary displacement of workers, but once the upheaval due to change settles down, those displaced employees almost always have new, usually better, jobs.

What jobs will robots actually create? In the first stage of the industrial robot revolution, humans will be needed to design, make, and maintain the robots. These workers include electromechanical engineers, technicians, designers, servicers, and even blue collar laborers to construct the robots. Today, many people are being introduced to the concept of robotics by hands-on experience with probots like Heath's HERO 1, designed to be a training device for students and workers. (Chapter 3 is devoted to HERO 1.) This exposure allows persons to feel more comfortable with robots, so there will be less turmoil when changes caused by industrial robots start taking place (Figure 2-2).

Figure 2-2 Probots can be excellent tools for business education. *(Courtesy of Heath Co.)*

After years of planning, making, and servicing robots, the second stage of the industrial robot revolution will take place, in which factories are built around the robot, rather than the robots being adapted to the factory. New robotics companies will grow, requiring secretaries, advertising personnel, salespeople, accountants, business managers, and others to manufacture the robots. In these growing years of robots, it is more likely that there will be a lack of labor, not a lack of employment. In fact, there will be such a demand for workers that the United States might find it necessary to open its borders to get help from abroad. There should be little worry about unemployment during this stage.

In these first two stages, a great deal of work will have been created. In the final stage, robots will be sophisticated enough to maintain themselves, requiring little human intervention. Will workers be needed from then on? Will humans be obsolete? Certainly not. It is inconceivable that the machine will ever rival the human mind. True enough, a computer can work with mathematical figures much faster than a person, but no computer has yet written a poem, thought of an idea, or drawn a picture completely of its own will. Probably the only way machines could ever approach true intelligence would be if a machine were somehow interfaced with the human mind—and even that doesn't make the machine genuinely smart, since the human is doing the thinking. Minds will always be needed, and they will be of even more use to us as our technology develops. We will all certainly have a great deal of leisure time to do the things we really enjoy, but we will always have the opportunity to think, create, and explore.

In the carefree life we hope to find in the third and final stage of the industrial robot revolution, it is suspected that we will all prosper. Robots will take care of food production, manufacturing, and nearly everything else humans have to manage today. Everyone will be wealthy since all of the necessities of life will be available. Of course, if everyone sat around waiting for robots to serve them, life would seem very meaningless. So, we can explore our own world, outer space, and ourselves more fully with our new tools and the great amount of time everyone will have. These are optimistic speculations about what this new age could be like. In reality it is very difficult to say what the industrial robot revolution will eventually bring us. The possibilities are, nevertheless, exciting.

THE HUMAN FACTOR

There is no doubt that industrial robots have the potential to do great things for our society. There will be troubles, though. Retraining workers, skilled and unskilled, for new tasks, is a major hurdle we must confront. Teaching people not to fear robots is another. More leisure time and new worlds to explore await us, but we will indeed have to pay the price.

It is hoped that any fears you may have had about the robots of industry have been reduced. Robots could continue to grow with us, increasing productivity and bringing better jobs for humans. On the other hand, we could condemn the robots as job thieves, continue to do work the way we always have, and let foreign manufacturers soar ahead of us with efficient and productive robots at their side. The choice is ours.

CHAPTER THREE HEROES ALWAYS WIN

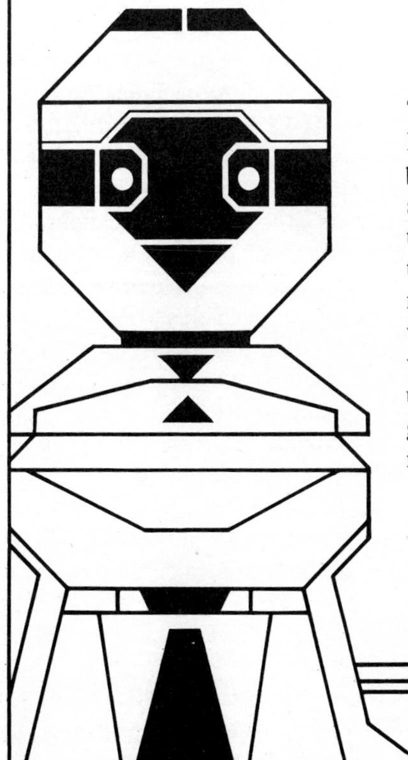

The Heath Company, located in Benton Harbor, Michigan, has something of a reputation for being a predictor of upcoming technology. In other words, when Heath comes out with a new product, it usually marks the beginning of a new technological revolution, be it big or small. The personal computer revolution which occurred around 1977 is a good example. Just before personal computers became popular, Heath had begun manufacturing personal computer kits, intended to educate electronics enthusiasts about computers and also to provide functional computers for the home. Heath computers are certainly not the most popular, but they did help spark interest in computing and provide a starting point for the personal computer revolution.

Heath's main products are electronics kits, some as simple as clocks and others as complicated as computers . . . or robots. The first probot to be produced by Heath is the HERO 1 robot (Figure 3-1). HERO is an acronym for Heath Educational RObot, which is exactly what it is: a robot designed to teach people about robotics, how to program a probot, and how to utilize a machine of this nature.

The HERO 1 comes in basically two forms—assembled and unassembled. The assembled form costs about $2,500 and requires only the most minor assembly, such as snapping on body panels. The assembled HERO 1 is the best product for those interested in learning about and using robots, but who don't feel very confident about soldering electronic components together or putting motors and microprocessors in their proper locations.

The unassembled HERO 1 costs only $1,500 but requires about eighty hours worth of work to complete. Soldering, assembling, and adjusting are all needed to finish this HERO 1, but Heath says the construction of the probot really isn't difficult for a person who knows how to solder well. Obviously, the kit is the more economical way to purchase a HERO 1.

The HERO 1 is a squatty-looking device weighing about thirty-nine pounds, measuring twenty inches high, and looking somewhat like the robot R2-D2 from the movie *Star Wars*. HERO's features are extensive. This probot can be programmed through a keypad for a variety of functions, including moving itself around, talking, and picking up objects with its arm. HERO can also rotate its head, sense light, detect sound, determine distance, and tell time.

This chapter is meant not only to teach you how to use a HERO 1, but also to illustrate how simple a probot can be to operate. Even if you don't own a HERO 1, this chapter will

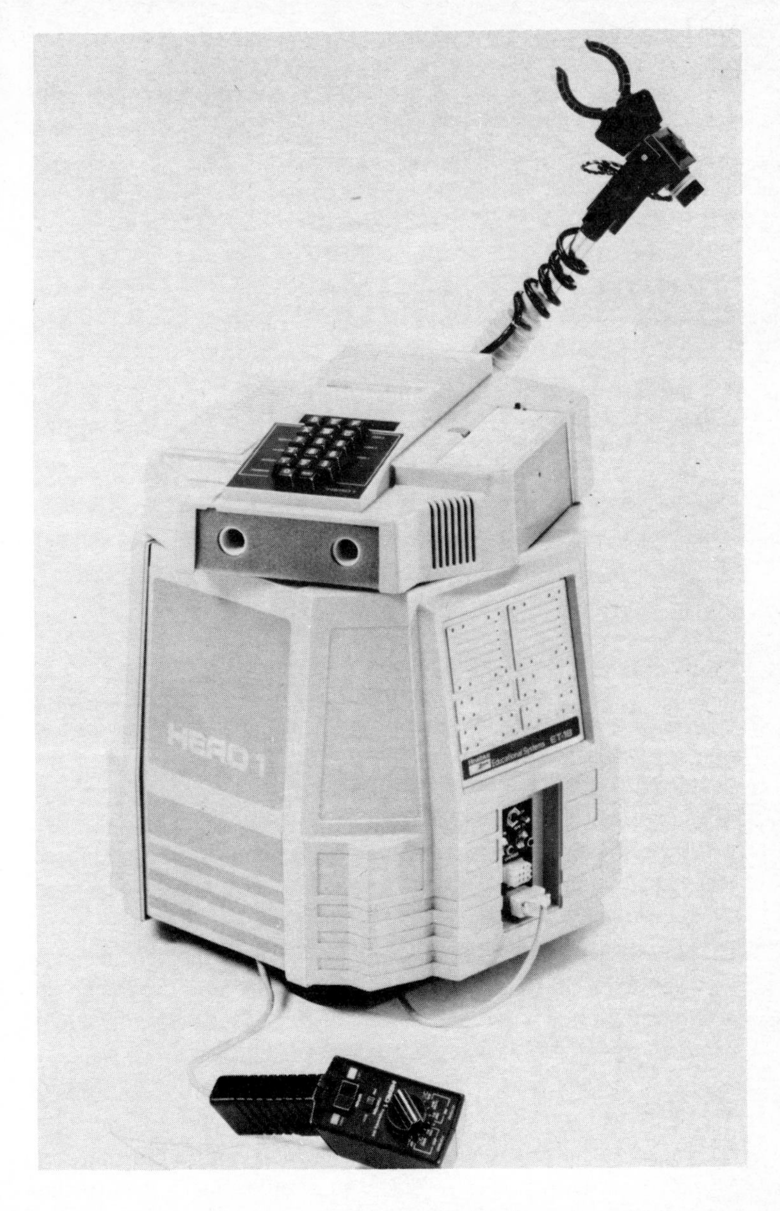

Figure 3-1 HERO 1, the Heath educational probot with its arm fully extended. *(Courtesy of Heath Co.)*

teach you much about the abilities of computers and robots. Finally, if you intend to buy a probot, you can read about the kinds available and then decide which one would best suit you. This chapter, then, can be an educational source, a shopping guide, or simply a compilation of interesting information about the HERO 1 probot.

BASIC FACTS ABOUT HERO 1

Before learning how to actually operate the HERO, it would be a good idea to have a basic understanding of the features of this probot. The programming needed to use these functions will be discussed later, but the facts are summarized below.

Senses

Sound: HERO 1 can detect sounds ranging in frequency from 300 to 5,000 hertz (cycles per second) with an accuracy of 99.996 percent. The probot cannot only detect the sound, but also can display the frequency to within approximately 19 hertz.

Light: The HERO can detect light within a span of thirty degrees in any direction of its light sensor. It can also detect the quantity of light with an accuracy of 1 part in 256, which is the same high accuracy as the sound sensor.

Motion: The motion sensor can detect any motion within fifteen feet, such as an adult walking or a person getting out of bed in the morning.

Ultrasonic Ranging: This probot can determine the distance between itself and another object up to eight feet away with an accuracy of plus or minus one centimeter. You probably won't hear the sound, but it could drive your dog crazy.

Speech: The HERO 1 is equipped with the popular Votrax SC-01 speech chip. This chip has sixty-four phonemes (sounds that can make up human speech patterns) which can be put together to make words, sentences, and even simple music. It also has the ability to inflect words for emphasis or sound effects.

Time: The electronic time circuit keeps track of seconds, minutes, hours, day of week, day of month, and month of year with an accuracy of plus or minus two minutes per year.

Head

The "head" of this probot houses all of the sensing devices and can be rotated up to 350 degrees. The head also contains a "breadboarding" area for electronics experimentation. This experimental breadboard provides direct access to an input/output port, computer control lines, a 12-volt power supply, a 5-volt power supply, and a user-defined interrupt mode for stopping the computer when certain conditions arise. The "head" also holds the hexadecimal keypad and the display.

Arm

The arm of HERO 1 can be rotated 350 degrees around and 150 degrees up and down. When retracted to its minimum length, it can lift up to a pound of payload. When the arm is extended (see Figure 3-1), however, it can lift only half that amount. It has a gripper at the end (for grasping objects) which may be opened up to five inches or closed completely. The gripper itself can be pivoted up and down a total of 180 degrees, while the "wrist" can be rotated up to 350 degrees.

Body/Base

The body houses the batteries of the probot and the three wheels which allow the HERO 1 to move (Figures 3–2, 3-3).

OPERATING HERO 1

Owning and using the HERO 1 is not a difficult task, though you will have to learn some things about programming before you can utilize its full power. When the HERO 1 is first turned on, it will say *"ready"* and show "HEro1.0" on its display. The

Figure 3-2 HERO 1 with its body panels removed.
(Courtesy of Heath Co.)

Figure 3-3 The "internal" HERO 1. *(Courtesy of Heath Co.)*

reason for the crude lettering is that the display consists only of seven eight-segment LEDs (Light Emitting Diodes), similar to those found on calculators.

When using the HERO, some precautions should be kept in mind to prevent it from being damaged. First of all, it should remain in a dry, clean place. If you send it into the rain or a dusty environment, it will almost certainly be damaged. Also, static electricity and deep-pile rugs are enemies of a probot. Lastly, the machine should be cleaned of dust and debris with a damp cloth, though water should not be sprayed on the probot.

There are several ways you can command the HERO 1. The most important method is through the keypad, located on top of its head. These keys are the best way to communicate with the probot, and many of them have more than one purpose. Almost all of the keys are numbers, consisting of the digits 0–9 and the letters A, B, C, D, E, and F. The reason for the letters is that the HERO is programmed using hexadecimal numbers. These are numbers in the Base 16 system, which is commonly used with computers. Therefore, the "number" after 9 is A, since A is equal to 10 in hexadecimal. Consequently, B is equal to 11, C is equal to 12, and so on. Understanding hexadecimal numbers is not terribly important when using the HERO 1, unless you want to experiment with the machine language of the probot's microprocessor. (Machine language will be dicussed later in this chapter.) To command HERO 1, it is only important that you understand what all of the letters mean.

The RESET key, located to the right of the zero key, tells the probot to go back into the *executive mode.* The executive mode is a *starting place* for the user of the HERO 1, since he can instruct the probot to go to any other mode (such as manual or learn) from this mode. The different types of modes will be discussed later in this chapter.

The ABORT key has the singular purpose of stopping whatever action in which the probot is engaged. This key is usually used when the HERO is doing something incorrectly and the user of the probot wants to correct the problem. When the ABORT key is pressed, the message "Abort" appears on the display. This number represents the last place in memory where an instruction to the probot was executed and is helpful in correcting the problem.

The sleep switch, though not a button, still serves an important purpose. The switch has two positions, Normal and Sleep. When pushed to *normal,* the probot behaves regularly. When the switch is pushed to *sleep,* however, the probot turns off power to everything except its memory, so that it can conserve energy yet still retain the program currently in its memory. Care should be taken with the sleep switch, though, since pressing the RESET key or turning the probot on again while the HERO 1 is sleeping will erase the entire contents of the probot's memory.

When the probot is in the executive mode, meant for *executing* commands directly from the keypad, you can tell the computer (the HERO is basically a computer on wheels) to go to another mode of operation. The executive mode commands are:

Key 1: go to the *program* mode
Key 3: enter the *utility* mode
Key 5: change to *manual* mode
Key 7: go to the *learn* mode
Key A: enter the *repeat* mode

As you type in commands, and as the computer is relaying information back to you, the display will be of great use to you. Even though it consists only of seven LEDs, it is still essential when programming and using the HERO 1.

THE UTILITY MODE

The *utility* mode is designed to help you as a programmer of the HERO 1. There are a number of commands in the utility mode, each serving a distinct useful purpose. These commands may be executed when you are in the utility mode, which you may enter by pressing the "3" button while in the executive mode.

The first utility command is for initializing the probot. This is accomplished by pressing the "1" key, which will return all of the HERO's components to their "home" position. When you initialize the probot, the following will happen:

1. The gripper will close completely.
2. The wrist will pivot to an upward position.
3. The gripper's wrist will rotate completely counterclockwise.
4. The arm will retract completely.
5. The shoulder will go down completely.
6. The head will return to its normal forward position.
7. The front wheel will turn all the way to the left.
8. The front wheel will return to the center position.
9. The computer will return to the executive mode.

Initializing the probot is useful since it has no way of knowing the positions of any of its components. By starting from these home positions, a HERO user can be assured that the programs made for the probot will function properly.

Another homing command can be invoked by pressing the "2" key for arm homing. This will return the arm, head, or steering to their home positions if any program leaves any of these components in a position which is different from the initialized position.

Since programs made for the HERO 1 take time and effort, it would be a waste to simply lose all of the programs made for this probot. For this reason, it has been made possible to transfer programs from HERO's memory onto a cassette tape for permanent storage and retrieval. Therefore, even though the program is lost when the HERO 1 is turned off, the program will still be on the cassette tape. To download—that is, save—a program from the probot's memory to cassette, the "3" key should be pressed. Once it has been pressed, the probot will ask the beginning and ending memory addresses of the program you wish to save. After you have specified these locations, the downloading process will begin, as long as you have a tape recorder connected to the computer by the cable provided with the HERO 1.

The reverse of downloading is uploading (program retrieval), which can be accomplished by pressing the "4" key. Once again, the cable should be hooked from the tape recorder to the probot so the program can be successfully loaded into its memory.

The commands to set the time, set the date, display the time, and display the date are the keys "5," "6," "7," and "8," respectively. When setting the time, you need only input the hours, minutes, seconds, and the status of the time (AM, PM, or twenty-four hour mode). For the date, the input should be in the format "year month day" so that a day such as August 9, 1983 would be input through the keypad as 830809.

THE MANUAL MODE

One of the easier ways to make the HERO 1 work is with the teaching pendant provided with the probot (Figure 3-4). This pendant has keys to actually control the probot while watching its actions. The *manual* mode, entered by pressing the "4" from the executive mode, is exclusively for the teaching pendant.

The trigger switch, like the "Fire" button on a game joystick, is the action button on the teaching pendant. Though you use the other switches to actually specify functions, the trigger switch is made for carrying out an action.

The function switch is a toggle switch for specifying whether you want to control the head and arm (*arm* mode) of HERO or the body (*body* mode). If the body position is selected, you can pick the probot drive operations: three speeds forward, three speeds backward, and neutral.

The rotary switch on the pendant controls the action that the head or arm motor will perform in the arm mode, or will select the direction and speed the probot should travel in the body mode.

Finally, the motion switch is used to tell the probot the direction the arm and head motors should run, or the direction the front wheel should steer. Although this switch is simply for specifying direction, it can be very helpful for certain purposes, such as rotating the probot completely around within its own diameter by using the HERO's steering.

The manual mode is an excellent way to begin understanding the features of this probot. It is also a good way to demonstrate the power of HERO, since the commands made through the teaching pendant are carried out by HERO almost immediately.

Figure 3-4 HERO 1 and the teaching pendant. *(Courtesy of Heath Co.)*

THE LEARN MODE

An extension of the powerful manual mode is the *learn* mode. Using the learn mode, you can control the probot's actions through the teaching pendant, and each action will be saved in its memory. Therefore, when you finish instructing HERO, every movement made will have been "remembered" as an actual program. The program itself can be saved for later use, or executed immediately so that you can see the probot in action.

Because of the learn mode's similarity to the manual mode, there is no need to explain again how the teaching pendant is used. There is, however, a specific way to enter the learn mode.

1. After the probot has been initialized and is in the executive mode, press the "7" key to enter the learn mode.
2. After the teaching pendant has been plugged into the HERO 1, the memory address where you want your program to start should be input through the keypad.
3. After the starting memory address has been input, the ending memory address should be typed into the HERO's keypad.

At this point, HERO is ready to "learn" what you want it to do. The teaching pendant is used as in the manual mode so you may program any movements desired into its memory. This makes programming the probot quick and simple.

If a mistake has been made while programming, you can back up as many steps as you like to correct the problem, just press the "B" key. The probot will then wait for you to press the trigger on the teaching pendant; as long as the trigger is pressed, the probot will continue backing through its program, step by step, with a brief pause between each step for you to release the trigger if desired. Once you have gone backwards to the place you want to change the program, press the "F" key. The probot will then accept your modifications. If, on the other hand, you decide the program shouldn't be changed at all, simply press the RESET key.

One other function of the learn mode is the RTI (Reverse The Instruction) feature. This will undo any actions that you have previously made. For example, if you made a program for the

HERO to move forward slowly, move its arm up, then turn its head counterclockwise, then you could use the RTI feature to make the HERO turn its head clockwise, move its arm down, and move slowly backward afterward. To do this, press RTI key "7" then hold down the trigger button on the pendant for the number of steps you want undone. If you simply wanted the HERO to reverse its last movement so that it would turn its head clockwise after turning it counterclockwise, you would hold the trigger button long enough for one command to be reversed. You can continue to hold the trigger button for as many actions as you want to be undone. The HERO will help, since it pauses briefly between the reversal of each instruction. Once you are through using the RTI feature, press the "F" key and enter any other steps in the program that you want.

The learn mode is probably the best way to begin programming the HERO 1, especially if you are unfamiliar with computers. In addition, it can provide an easy way to program the probot for specific tasks which can be saved on cassette tape for future use.

THE REPEAT MODE

The *repeat* mode is most commonly used to RUN or execute a program. From the executive mode, the repeat mode is entered by pressing the "A" key. Then the "D" key is pressed to instruct the probot to execute—or do—a program. The memory address from which you would like to start should then be typed into the keyboard. The HERO 1 will begin executing whatever is found at that memory location, and can be stopped by two different means:

1. Pressing the RESET key, which returns the probot to its executive mode with the program still in memory.
2. Pressing the ABORT button which will stop the program and show the value of the PC (program counter) on the display. The program counter simply tells at what place in the probot's program memory an instruction is currently being executed. This is helpful in case there is a problem with the program, since you will know exactly where the error exists.

THE PROGRAM MODE

The *program* mode, entered from the executive mode by pressing the "1" key, is the most intricate, most versatile, and most educational way to program the HERO 1. Unfortunately, however, it requires that the programmer know the language of the 6800 microprocessor by which the HERO runs. The language of the microprocessor, called machine language, is input into the computer strictly through hexadecimal numbers (such as F4, B6, and 1E). Though it is fast and efficient, it requires a great deal of forethought and work on the part of the programmer.

Fortunately, there is another program mode available which makes programming somewhat easier. Although it still uses all of the machine language instructions that the regular program mode uses, this second, *Auto*, mode contains additional commands which perform specific tasks. This is called an *interpreter language,* since the HERO must interpret these special codes (see Table 3-1) and translate them into its lengthy and often difficult machine language. Because the probot must interpret these special instructions, it can take from 10 to 100 times longer to execute a certain command. However, since the microprocessor controlling the commands works much faster than the motors controlling the probot, there will be almost no difference in time when the program is actually executed.

Table 3-1 Interpreter Commands

Command to Computer	Action
02	Stop drive motor
03	Stop steering motor
04	Stop arm motors
05	Stop speaking
3A	Return to executive mode
41	Enable light detector
42	Enable sound detector
45	Enable ultrasonic ranging
4B	Enable motion detector
72	Speak
83	Change to machine language
87	Go to sleep

The command ''83'' tells the probot to go from its interpreter into machine language. You can instruct the probot to go back to the interpreter with the machine language code 3F.

The memory of the probot is addressed by hexadecimal numbers, ranging from the bottom of memory (0000) to the top of memory (FFFF). The area allocated for your use resides from 003D to 0EE7, which makes 3,754 bytes available to you. A byte is a unit of information (usually a character of information) in which you can store an instruction, so even if you had a program with a full 375 instructions, you would have used only about 10 percent of the memory available to you.

Although a detailed explanation of machine language cannot be given here, there are several good books on the 6800 microprocessor which may be found in the computer section at any large bookstore. In addition, the manuals supplied with HERO contain the machine language commands used with the microprocessor. You should know, however, that machine language is sometimes tedious and frustrating, and patience is required to master it. For this reason, other ways of programming the probot (such as the teaching pendant and the interpreter) have been provided with the HERO 1.

Rest assured, entering a program into the probot is not a difficult task. Assuming that the auto mode is used, since it offers everything the program mode does and quite a bit more, here are the steps that must be taken:

1. Press the ''A'' key while in the executive mode two times to instruct the computer that you want to enter a program with the memory addresses numbered automatically.
2. Enter the four-digit hexadecimal number specifying where you want your program to begin in HERO's memory.
3. Begin entering the machine language program (or the interpreter commands), one command at a time. The memory address where you are currently located is displayed in the left four digits of the display, while the two hexadecimal digits you enter as your command are displayed in the two right digits. If you make a mistake, you can correct it when you have finished the program. It is a good idea, however, to make a note of where in the memory the mistake was made.

Once the program has been finished, press the RESET key. From that point, you can run the program, save it on tape, or modify it. To modify the program, press the "A" key again, then the "E" (for examine) key, followed by the memory address you wish to examine. You may use the "F" and "B" keys to scroll forward and backward through memory, and you may press the "C" key to change any instruction. If the "C" key is pressed, the two hexadecimal digits will go blank, and the probot will wait for you to enter a new instruction. When you are through modifying the program, simply press the RESET key.

When you are ready to run—that is, execute—a program, you need only press "A," "D," then the memory address where your program begins.

To help you understand what a program would actually contain, here is a sample program (Table 3-2) in the HERO 1's interpreter language. This short program instructs the probot to turn its head around, pause, go forward, pause, turn its head forward again, pause, back up to its original position, pause, then return to the executive mode. Keep in mind that this program is slower than machine language, but a machine language program would be even longer than this one.

Once you review the program, you may wonder why some commands appear to do the same thing, but consist of different numbers, such as the D0 and D4 instructions, both of which move the head motor. Actually, D0 moves the head motor to the right, while D4 moves the head motor to the left.

As this sample shows, making programs isn't as difficult as it might seem. Programming the probot is challenging and educational, especially when using machine language, and you'll soon find that your experience is the greatest teacher.

MAKING HERO 1 TALK

One of the HERO's most exciting and useful features is its voice chip, which allows it to "say" sixty-four different phonemes, or basic sounds of speech. When the phonemes are used properly, words, sentences, and sound effects are possible. Even though

Table 3-2 A Sample Program for the HERO 1

Memory Address	Instruction	Purpose
0100	D3	Drive a motor
0101	D0	Use head motor
0102	60	Right 60 units
0103	BD	Go to subroutine
0104	01	at address 0119
0105	19	---(*pause*)---
0106	D3	Drive a motor
0107	10	Use wheel motor
0108	10	Forward 10 units
0109	BD	Go to subroutine
010A	01	at address 0119
010B	19	---(*pause*)---
010C	D3	Drive a motor
010D	D4	Use head motor
010E	60	Left 60 units
010F	BD	Go to subroutine
0110	01	at address 0119
0111	19	---(*pause*)---
0112	D3	Drive a motor
0113	14	Use wheel motor
0114	10	Backward 10 units
0115	BD	Go to subroutine
0116	01	at address 0119
0117	19	---(*pause*)---
0118	3A	Executive mode
0119	CE	Put the number
011A	01	0100 into the
011B	00	index register
011C	09	Decrement index
011D	26	Is index=0?
011E	FD	If not, keep subtracting
011F	39	Return from subroutine

the voice doesn't sound as realistic as yours or mine, it is still understandable.

The phonemes are stored as hexadecimal numbers within the computer's memory. Each phoneme makes a distinct sound. For example, the phoneme SH (code number 11) makes a sound like the beginning of the word "*sh*irt." To make the entire word "shirt," four phonemes must be used, SH ER R T (11 3A 2B 2A).

A simple program to make the probot talk is shown in Table 3-3. To add your own speech, merely type in the phonemes you want beginning at memory address 0095. You may add as many phonemes as you like, since it does not matter where the program ends, as long as you don't run out of memory.

The HERO comes with a voice dictionary so you can program more complex words and sentences without having to figure out the phonemes yourself. The phonemes resemble English, so they are not too difficult to decipher. For example, R 01 U1 B AH1 UH3 T Z L UH3 AH2 Y K M Y1 IU U1 U1 Z I1 K is "robots like music," as one could find out by sounding out each of the phonemes.

The HERO's voice can be inflected one, two, or three levels. Inflection makes the HERO's voice sound more forceful, so that words may be emphasized or singing may be simulated. To inflect a phoneme, add the hexadecimal digit 40 to the phoneme's code. Therefore, the phoneme 01 would be 41 if inflected to the first level, 81 if inflected to the second level, and C1 if inflected in the third and highest level. Using phonemes, you could make the HERO say such things as, "I'm not going to let *you* touch my keyboard," or "*Halt* burglar, or I'll *blast* you!" Finally,

Table 3-3 The Talking Program

Memory Address	Instruction	Purpose
0090	72	Say phonemes
0091	00	that begin at
0092	95	memory 0095
0093	20	Wait here when
0094	FE	talk is over
0095	11	"SH" (sound)
0096	3A	"ER" (sound)
0097	2B	"R" (sound)
0098	2A	"T" (sound)
0099	3F	Stop talking
009A	FF	End of phonemes
009B	3A	Executive mode

sound effects can be created by repeating certain phonemes over and over. For example, you can create the sound of a laser by using the phonemes D (1E) and J (1A) repeatedly.

The voice has two mechanical controls which can be used to change the quality of HERO's voice. The volume control can be set to change the loudness of the phonemes, while the speed/pitch control can control how quickly HERO talks. In addition, the pitch of the voice can be controlled, for the quicker he talks, the squeakier he sounds.

THE POTENTIAL OF HERO 1

There are indeed many possible uses for the features and functions of HERO 1. This probot's arm, voice, sensors, ability to move, and memory give it the power to execute many tasks, some of which will be discussed in Chapter 6. For now, though, here are a few possibilities for you to ponder:

1. *Using the HERO as a "Babysitter."* It could make sure the children don't turn the television volume up too loud and could tell them when it is time to go to bed. After the children are asleep and not moving, the HERO could turn off the lights.
2. *Programming the HERO to Respond to Different Frequencies.* A low pitch could make HERO walk around the room and remark how much it enjoys listening to bass music, while a high-pitched whine could make it whirl its arm in different directions, asking for the shrillness to be stopped. This would be a good demonstration of the probot's capabilities.
2. *Instructing the Probot to Stand Guard at Night, Patrolling Different Parts of the House.* If a motion were detected, the HERO could shout, "HALT! Who goes there?" If it detected a particular response, such as three claps, it could return to its patrolling. If, however, the specific response were not detected, the HERO could yell "ALERT! ALERT! INTRUDER SPOTTED," and race around the house informing the occupants about the intruder. This would probably be unnecessary, though, since the would-be burglar would most likely flee in terror.

The HERO 1 is a powerful and useful probot of today, and assembling the kit form can give you an even greater understanding of how a robot works. Regardless of whether you purchase a kit or an assembled model, the HERO should prove to be a versatile and educational companion.

There are rumors that a HERO 2 will be on the market before long, with all of the abilities of the HERO 1 along with voice recognition, shape recognition, and wireless communication with other electronic devices. The HERO 2 would be an exciting addition to the Heath family, but for now, the HERO 1 is an excellent probot with a lot of potential as a household helper and useful companion.

CHAPTER FOUR

MY THREE PROBOTS

Nolan Bushnell, one of the most famous entrepreneurs in the country, was a very important figure in the personal computer revolution. In 1972, Nolan and his friends created a game called Pong in a garage located in the San Francisco Bay area, later to become the famous Silicon Valley.

Many large companies thought that Pong was a game with little promise. These corporations believed that video games were never going to be a significant entertainment industry, so they ignored Bushnell's idea. However, Pong became extraordinarily popular, and Nolan's new company, Atari, was destined to become one of the largest and most powerful makers of consumer electronics products.

Atari grew, just as Bushnell's other ventures did, and video games continued to become more sophisticated and exciting. Nolan formed other industries such as Chuck E. Cheese's Pizza Time Theatre, Catalyst Technologies, and Androbot, Inc. The products of Androbot will be our main interest for the next two chapters, though, since that company is one of the most promising and productive makers of personal robots.

THE ANDROID ROBOT

The name Androbot is derived from two different words: android and robot. We are already familiar with what a robot is; an android is simply a mechanical device which resembles a human. Therefore, an Androbot is a robot that looks in some ways like a human being. This is an advantage for the company. Robots resembling people are likely to be the most popular probots, since humans tend to be more receptive to their own kind. If a person were placed in a room with a squatty-looking probot and a friendly, human-like probot, that person would probably be more inclined to use the android. In a word, androids are thought by many people to look cuter than other less attractive robots, such as the industrial variety.

Androbot, Inc. currently makes four different types of probots: AndroMan, F.R.E.D., Topo, and B.O.B. (Figure 4-1). The first three will be covered in this chapter, while B.O.B. will be discussed in Chapter 5. These personal robots are good-looking machines as well as useful tools, so our exploration into the potential of the Androbots should be interesting.

Figure 4-1 Nolan Bushnell and the Androbot family. *(Courtesy of Androbot, Inc.)*

ANDROMAN

The smallest and least expensive Androbot is AndroMan (Figure 4-2), which is basically designed to play games with a human through the Atari 2600 video computer system. This is a big step both for games and robotics, since the AndroMan introduces a new type of video game—real-life action. Never before has an actual object been employed with the Atari 2600, but with AndroMan many exciting games are possible, and more interest in robotics will probably be generated.

This twelve-inch high probot, controlled by a joystick via a infrared signal, moves about on a three-dimensional playing surface while the game action occurs on the video screen. The AndroMan is a supplement to the game, since two areas of action exist—on the television screen and on AndroMan's three-dimensional playing surface. This probot also comes with a game cartridge, a set of game pieces imprinted with coded information, and an instruction manual.

The idea behind the AndroMan games is that there are two different scenarios of game play: the monitor screen and AndroMan's "domain." After you have accumulated a number of points by playing the video game, you must guide AndroMan around the 3-D landscape to intercept coded game pieces which will activate a different type of game play. AndroMan even speaks to the player, sometimes congratulating him for excellent performance, sometimes warning him of impending danger.

Androbot will be coming out with new game cartridges and AndroMan accessories in the future, and with more than ten million Atari VCS units already existing, it is very likely that AndroMan will be a successful new probot. Although this is a robot in its simplest form, it still provides a unique and flexible enhancement to home entertainment, since the element of a real-life talking object is introduced to the video game that used to exist only on the screen.

F.R.E.D.

The acronym F.R.E.D. stands for Androbot's Friendly Robotic Educational Device (Figure 4-3). This twelve-inch high probot

Figure 4-2 AndroMan in action. *(Courtesy of Androbot, Inc.)*

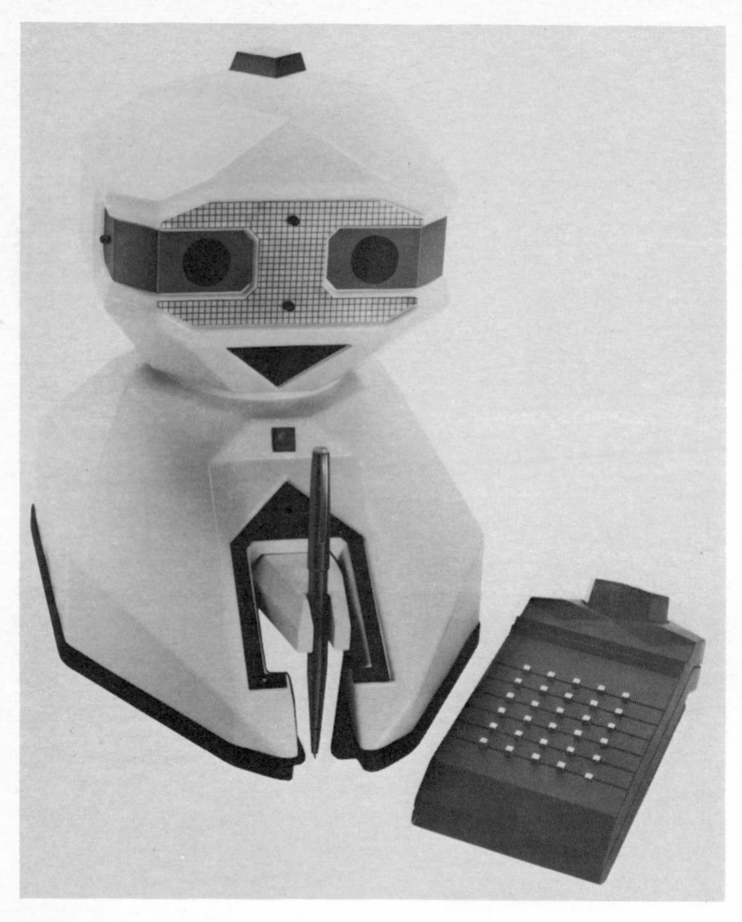

Figure 4-3 F.R.E.D. *(Courtesy of Androbot, Inc.)*

will probably be one of the most popular for young people, since its low price and educational theme are attractive to many parents who want their children to know more about computers and robots.

This probot is designed to be an extension of a personal computer, such as the Apple II. F.R.E.D. may be controlled without a computer, however, through the infrared controller provided with it. F.R.E.D. also has sensors of its own, so it won't fall off the edges of tables or stairs, both of which would probably damage or destroy its electronic components.

F.R.E.D. looks like a human head mounted on an ivory-colored block, and its potential lies mainly in its ability to move about. This probot can move in any direction and can remember the patterns in which it moves for future reference. Also, a pen may be attached to its base so F.R.E.D. can draw pictures (geometric or otherwise) on a piece of paper.

F.R.E.D. also comes with a small wagon for carrying around tiny payloads. This probot is equipped with a voice that has a forty-five word vocabulary, which can be used while it is being controlled by a computer. Software packages are available for F.R.E.D., and Androbot reports that accessories, such as a robotic arm, will be available soon.

The F.R.E.D. is a good introduction to robotics, since the machine is low-cost and can be interfaced with a computer easily. However, for a person wanting a probot with a lot of potential, the F.R.E.D. might not be the best bet. Its abilities are basically limited to moving around, talking, and drawing shapes. Although this is a fine educational tool and a great way to learn more about robots and computers, those interested in a full-fledged probot might want to consider something more advanced—and expensive.

TOPO

Androbot calls Topo a mobile extension of your personal computer. As that implies, Topo requires a personal computer in order to operate. When introduced, Topo could only work with the Apple II computer, but Androbot promises versions for the Commodore, Atari, Texas Instruments, and IBM personal computers in the very near future.

The word *Topo* is short for topology, which is what this probot excels in, since it can be manuevered around an environment and made to remember the layout of that environment, such as a house. For example, if you guided Topo through your home once, it would be able to recall the layout of the entire house and move itself to any room without bumping into walls or other objects. This is an extremely useful feature, since you could instruct Topo to ''Return to the living room,'' it would know immediately how to get to the room without encountering any problems.

Topo is an attractive probot for several reasons:

1. *Its Low Price.* Topo costs about $795 without a voice, or $995 equipped with a voice.
2. *Its Size.* Unlike its shorter, younger brothers, Topo is over three feet high and looks like a genuine robot. It does indeed have something like the shape of a human, complete with head, torso, and sometime in the future, arms.
3. *It's Versatility.* Since this probot is connected to a personal computer, it is limited only by the computer's capabilities. Topo can be made to perform many different tasks with Androbot's different software packages. It can even teach you more about computers, since they use unique computer languages such as TopoForth, TopoBASIC, and TopoLogo which must be learned in order to control the probot.

Topo is a particularly useful member of the Androbot family since it can move itself and cargo around at a speed of two feet per second with its versatile Andromotion. Andromotion is possible with the two angled wheels at Topo's base which allow it to move, turn, or rotate in any direction without falling over. Also, a voice-equipped Topo gives this probot the power to communicate with humans for the purpose of education, babysitting, or even entertaining party guests. Topo also has a panic switch near the top of its head. If Topo starts doing something you don't want it to do, a tap on the head will turn if off.

Maintaining Topo doesn't require much effort; an occasional

cleaning of the outside with a damp cloth makes up the majority of the maintenance. The drive wheels should be checked periodically to insure that no carpet lint is gathering on the wheels or wheel axle, and his environment should not be too hot, cold, humid, dusty, or electrically charged. Finally, Topo should be fully charged in order to work properly.

The Topo package for the Apple II computer includes:

1. The Topo probot
2. A manual and owner registration cards
3. Topo's battery charger
4. A foam block on which to prop Topo while calibrating its wheels
5. The Topo/Apple transmitter card to be plugged into the back of the Apple II
6. A radio transmitter for sending commands to Topo without any wires attached to it or the computer
7. A floppy disk containing a program to run Topo with a joystick and a set of BASIC Topo controller routines

PROGRAMMING TOPO WITH TOPOBASIC

One of the easiest ways to control this probot is through *TopoBASIC*, provided with the machine. Nearly everyone who owns a computer is familiar with the BASIC language, and the job of utilizing Topo's abilities is made much easier when TopoBASIC is employed.

Before using TopoBASIC, though, you must become familiar with how to move Topo. The joystick program is perfect for this. Once the diskette has loaded the Topo programs into the computer, by typing **GOSUB** 6000:**GOSUB** 5000, you instruct the computer that you want to use a joystick to move Topo around. If the joystick is pushed forward, Topo moves forward; if the joystick is pushed to the right, Topo steers to the right, and so on.

There are routines, however, within the TopoBASIC program that will allow Topo to run by itself. These BASIC subroutines may be accessed by you as many times as you want, to carry out specific tasks. Each of these routines is located at a specific line number (see Table 4-1), so a **GOSUB** statement, followed by a line number, will execute the subroutine located at the line number. These routines may be used individually, of course, but Topo can do more complex tasks when the subroutines are combined.

The motion commands are dependent on the variables you give them. The particular variable is called "N," and it is used to specify the number of steps that you want the probot to execute. For forward and backward motion, a step is approximately one centimeter. For steering the probot, a step is about one degreee. Therefore, if you wanted Topo to move a meter forward, you would type: N=100:**GOSUB** 5100. Or, if you wanted to turn Topo forty-five degrees to the right, you would type: N=45:**GOSUB** 5400.

Incidently, before using any of these commands, you should

Table 4-1 Subroutines for Topo

Line Number	Function
5000	Make joystick control Topo
5100	Move Topo forward
5200	Move Topo backward
5300	Turn Topo left
5400	Turn Topo right
5500	Stop Topo
5600	Move Topo motors specific amount
5700	Move Topo motors, then stop
5800	Delay
5900	Set up values for Topo's motors
6000	Clear all values for Topo's motors
6300	Store values for Topo's motor
6450	Store how many steps for each motor
6500	Move forward precise amount
6600	Move backward precise amount
6700	Steer left at precise angle
6800	Steer right at precise angle
7000	Initialize values for all motors
7300	Calibrate Topo's wheels

type in **GOSUB** 6000:**GOSUB** 7000, to reset Topo. This will access the initializing routines. If you write a program, you only need to type these commands once so that you begin with a "fresh slate."

With this in mind, we can write our first program. Remember that even if you don't have a Topo, reading this can still be helpful to you, since you can determine if this probot has the capabilities and simplicity of operation you're looking for. The program below simply moves the probot backward thirty centimeters, then forward thirty centimeters. It proves how simple TopoBASIC is to use.

```
10  GOSUB 6000:GOSUB 7000: REM INITIALIZE TOPO
20  N=30: REM MAKE MOTION 30 STEPS
30  GOSUB 5200: REM MOVE BACKWARD
40  GOSUB 5100: REM MOVE FORWARD
50  END
```

And it shows how easy programming a probot can be. Granted, there are more difficult and more efficient ways to program a probot (such as operating HERO 1), but for those learning about computers and robots, the BASIC language is an excellent way to begin.

Given the large variety of subroutines provided with TopoBASIC, complex movements are possible with this probot, as the following program proves:

```
10  GOSUB 6000: GOSUB 7000: REM INITIALIZE TOPO
20  LN=40: REM MAKE TOPO MOVE 40 STEPS
30  AN=120: REM MAKE TOPO STEER 120 DEGREES
40  FOR I=1 TO 3: REM DO THIS THREE TIMES
50  N=LN: GOSUB 5100: REM MOVE TOPO FORWARD
60  N=AN: GOSUB 5400: REM STEER TOPO TO THE
    RIGHT
70  NEXT I: REM CONTINUE LOOP UNTIL DONE
80  END
```

This program makes Topo follow the path of a triangle with sides of forty centimeters each. If you want the triangle to be larger, simply change the LN (length) value. If you want a different shape, change the angle (AN) value and the **FOR/NEXT** variable "I" to equal the number of sides of the shape.

For example, a square would have angles of ninety degrees (AN=90) and four sides (**FOR** I = 1 TO 4).

We have examined how forward motion, backward motion, and steering work on the Topo. There are still other commands that make this probot's motion even more versatile. These subroutines control the speed of Topo's individual wheels, which will allow you to make Topo move in curves as well as the sharp angles that we have already learned. The variables used with these subroutines are:

C1 = Left wheel motion
C2 = Right wheel motion
STP = Step value to determine rate at which Topo moves

This first sample program makes the left wheel move faster than the right wheel, causing Topo to create a circle in a clockwise direction:

```
1Ø   GOSUB 6ØØØ: GOSUB 7ØØØ: REM INITIALIZE TOPO
2Ø   C1=15ØØ:C2=125Ø: REM SET VALUES
3Ø   GOSUB 59ØØ: REM GIVE VALUES TO MOTORS
4Ø   END
```

This program will make Topo run until the command **GOSUB** 5500 is given to the computer which causes Topo's wheel motors to stop running. The step value (STP) may be used to slow down the probot after a shape has been drawn. Each STP value will delay the motion of the probot 1/50th of a second, so if STP is equal to fifty, there will be a one-second delay when the probot finishes a pattern, such as the circle created by the following program:

```
10   GOSUB 6000: GOSUB 7000: REM INITIALIZE TOPO
20   STP=500:C1=1500:C2=1250: REM ESTABLISH
     STEP AND MOTOR VALUES
30   GOSUB 5700: REM MOVE TOPO IN A CIRCLE,
     THEN STOP
```

Given the variety of subroutines that can manipulate Topo, your imagination is the only limit to this probot's motions. Accuracy is important, though, and that is the reason for calibrating this Androbot.

CALIBRATING TOPO

Topo's Andromotion is an excellent system of movement, but it is not perfectly accurate. For this reason, the wheels on Topo have adjustable speeds so that the probot will move properly. When it is told to move in a straight line, it should move in a straight line, not a diagonal one. When Topo is instructed to turn left ninety degrees, it should make a perfect right angle, and not turn left eighty-five degrees.

Making the desired adjustments in Topo involves two simple steps:

1. Finding the zero setting for the motors—that is, adjusting the wheel drives so that they move at the same rate
2. Calibrating the angle at which the probot turns and the distance it moves

The first step, finding the zero setting for the motors, is the easiest. To begin with, the lines 7005 and 7010 in TopoBASIC should be examined. They should look like this:

```
7005   Z1=1024
7010   Z2=1024
```

These variables, Z1 and Z2, are the settings for each wheel's drive motor. These variables must be set equal to the value "A" so that the calibration routine, supplied with TopoBASIC, will be able to determine the correct values for Z1 and Z2. Therefore, line 7005 should be changed to "Z1=A." To begin the calibration routine, put the Topo on the prop that comes with it, and type into the computer: **GOSUB** 7200.

At this point, the lights on Topo's left wheel will tell you how you should adjust Topo for proper calibration, and the calibration program will ask you what number should be used for that wheel. The first number to try, of course, would be the original value of 1024. If both lights are off or both lights are flickering, no calibration is needed for that wheel. However, if only the front light is on, input a slightly smaller number into the calibration program, such as 1021. If the back light is on, input a slightly larger number into the calibration program, such as 1027. Once the lights show the calibration to be correct, write down the number you used for that wheel. Go through the same process with the other wheel, using line 7010 and variable Z2 this time, and write down the number. Once you are through, you can input the two numbers into lines 7005 and 7010. For instance, if you found that the left wheel was calibrated with number 1020 and the right wheel worked properly with number 1028, you would enter:

```
7ØØ5  Z1=1Ø2Ø
7Ø1Ø  Z2=1Ø28
```

That's all there is to zeroing Topo's wheels.

Now the distance Topo moves and the angle at which it turns must be calibrated. To make sure Topo will move in a straight line, type into the computer: **GOSUB** 7300. You will now be asked to give a number for the speed of each wheel. For now, give the computer the number 1536 for both the right and left wheels. If Topo moves in a straight line, no changes need to be made. If one wheel goes slightly faster than the other, causing Topo to veer from his straight course, decrease the value of the faster wheel. Keep decreasing the value for the faster wheel until Topo goes in a straight path. Once you have the numbers for both the right and left wheels, set the variable F1 in line 7025 equal to the left wheel value and the variable F2 in line 7030 equal to the right wheel value. Therefore, if you had the numbers 1536 for the left wheel and 1475 for the right wheel, you would type:

```
7Ø25  F1=1536
7Ø3Ø  F2=1475
```

Now the distance that Topo moves should be adjusted. This is controlled by the SFD variable in line 7085 of TopoBASIC. Type in the following test program:

```
1Ø  GOSUB 6ØØØ:GOSUB 7ØØØ: REM INITIALIZE TOPO
2Ø  N=1ØØ: REM MAKE TOPO MOVE 1ØØ UNITS
3Ø  GOSUB 51ØØ: REM BEGIN MOTION
4Ø  END
```

If Topo moved more than one meter, decrease the value of SFD in line 7085. If Topo moved less than a meter, increase the value of SFD in line 7085. Continue running the same program until Topo travels exactly one meter.

Finally, the angle at which Topo turns should be adjusted. The left-turn angle is adjusted with the SLT variable in line 7095 of TopoBASIC. Run this sample program to see if SLT is set at the proper value:

```
1Ø  GOSUB 6ØØØ:GOSUB 7ØØØ: REM INITIALIZE TOPO
2Ø  N=36Ø: REM 36Ø DEGREE ANGLE
3Ø  GOSUB 53ØØ: REM TURN LEFT
4Ø  END
```

Topo should have turned completely around. If it turns a little more than a full circle, decrease the value of SLT in line 7095. If Topo doesn't quite complete a circle, increase the value of SLT in line 7095. Once you are through calibrating the left angle, calibrate the right angle by changing line 30 in your test program to **GOSUB** 5400 and adjusting the variable SRT in line 7100 as needed.

After all of these calibrations have been finished, erase the test program (lines 10 through 40) and save this revised version of TopoBASIC on your diskette with the command: **SAVE** TOPOBASIC. This will ensure that you have the proper calibrated values each time you use Topo.

THE USEFUL PROBOT

Topo's movement and optional voice make it an excellent extension of the personal computer. Though this probot is not autonomous like its older brother, B.O.B., or Heath's HERO 1 probot, it is an attractive and powerful machine.

Quite a few people are already using Topo in different ways, and Androbot has reported some of the more unique activities with which Topo is involved.

At Florida A & M University, Topo is used in research on how probots can be used to help disabled people. At California Lutheran College, students are using Topo to master the personal computer and to solve problems in using probots. An elementary school in California uses Topo to help young children feel more at ease with probots, and a "Topo Olympics" has already been designed for the probots to find their way through an obstacle course. Finally, one Topo in Minnesota was programmed to dance the hokey-pokey at a science exhibit. Five other students competed in the dance competition, but the probot won, of course.

Topo can be used for other purposes, naturally, whether they are educational, useful, or entertaining. One major benefit of Topo is that a user of the probot can become much more familiar with the language of the personal computer. In fact, TopoBASIC is not the only language you can master by using this probot. TopoFORTH and TopoLOGO are languages that are easy to use, which make them ideal for children and adults alike. The commands not only move the probot about, but also display the motion on the screen (making interesting designs). Graphics are drawn on the screen in the same direction as Topo moves, which can be useful for seeing the motion of the probot while it is in another room. Also, simple commands control the motion of the probot; for instance, FORWARD 50 RIGHT 90 moves the probot forward fifty centimeters and then turns Topo to the right ninety degrees.

TOPO AND THE FUTURE

Soon there will be more products for the Topo (such as robotic arms), more software, and consequently more uses. Topo is an excellent value, and an economical introduction to the world of personal robots. Topo, and the rest of the members of its Androbot family, will quite possibly be with us for some time to come.

CHAPTER FIVE
B.O.B., RB5X, AND MAKING PROBOTS

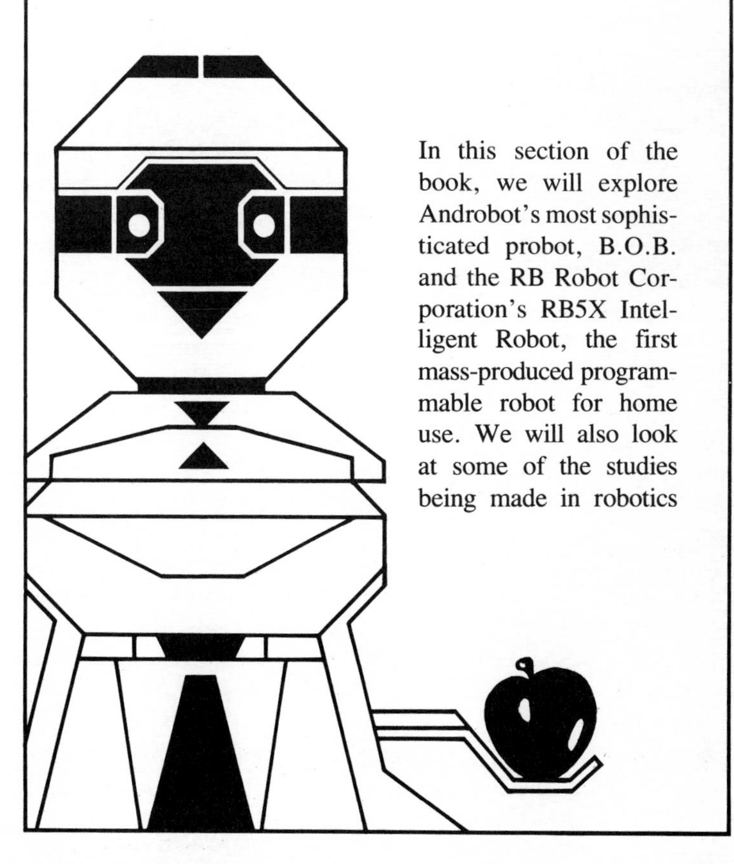

In this section of the book, we will explore Androbot's most sophisticated probot, B.O.B. and the RB Robot Corporation's RB5X Intelligent Robot, the first mass-produced programmable robot for home use. We will also look at some of the studies being made in robotics that could lead to more sophisticated technology for the probots of the near future. This will not only illustrate some of the things we can expect from probots in the next few years, but will also provide some insight about the increasing amount of work and energy being invested in robotics.

B.O.B., BRAINS OR BRAWN?

Androbot's B.O.B. is their most expensive and powerful probot (Figure 5-1). B.O.B. is an acronym for Brains On Board, since the B.O.B. is controlled by its own computer which is, as most computers are, printed on a circuit board.

B.O.B.'s greatest advantage is that this probot is completely autonomous. That is, B.O.B. doesn't require a personal computer to work properly, as Topo and F.R.E.D. do. In fact, B.O.B. has quite a substantial computer of its own. Because B.O.B., like the HERO 1, is entirely autonomous, it is not restricted to the confines of a home or an office where a computer and infrared transmitter would have to be located. B.O.B. can go just about anywhere and do practically any task within the capabilities of its three-foot high body and "mind."

B.O.B. was designed by Androbot to entertain, communicate, and to help around the home. This probot looks very much like Topo, except for a few extra lights on the front of the torso. But the brain of B.O.B. is the outstanding feature of this machine. B.O.B. is run by two Intel 8086 microprocessors, much like the microprocessors that run home computers. However, home computers usually have *one* of these chips, which goes to show how smart this probot is. B.O.B. also has a tremendous memory capacity of three megabytes, or twenty-four million pieces of information. This is about two hundred times as much memory as an average home computer, and, because of the sharply lowered price of the memory chip, B.O.B. is still a relatively affordable probot at $2,995.

B.O.B., like Topo, can move itself at a speed of two feet per second. He can also remember the pattern of a house so he won't bump into walls or obstacles. Also, this robot has a human-

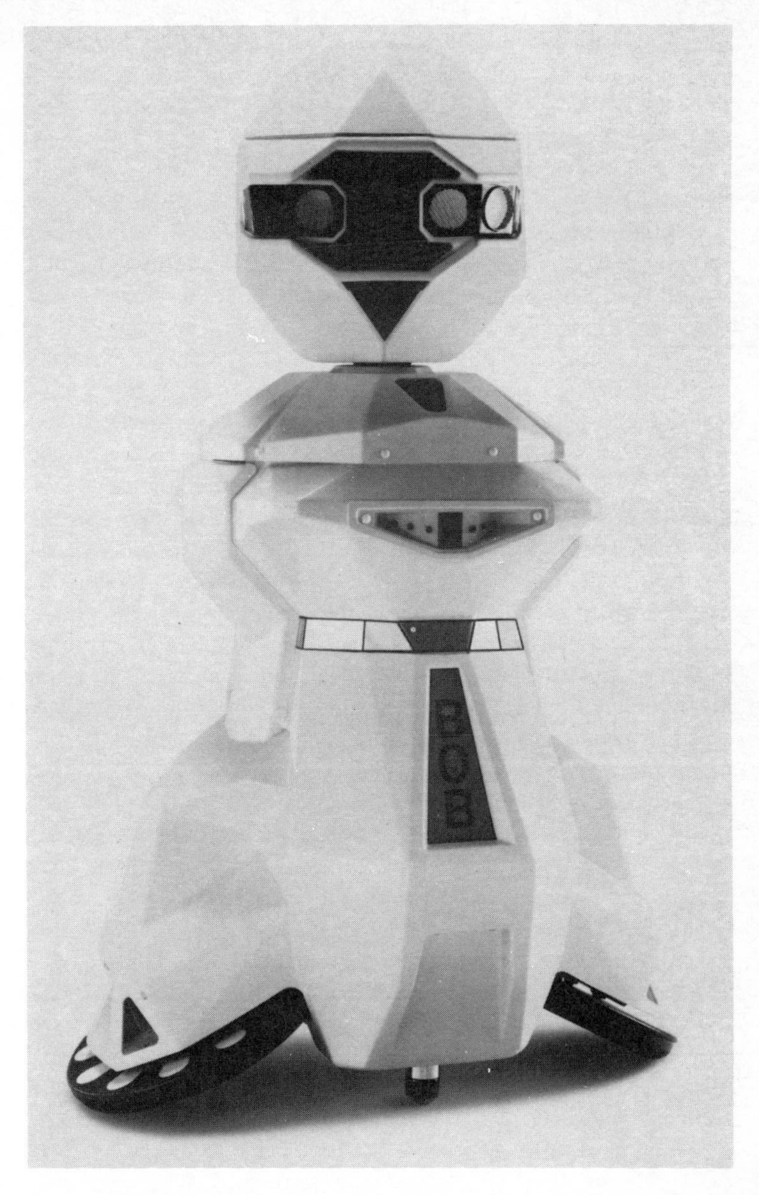

Figure 5-1 B.O.B. *(Courtesy of Androbot, Inc.)*

like voice which can recite words and phrases for many purposes. Sensors are also standard on B.O.B., which enables the probot to detect human beings (with a heat detector) and avoid obstacles (with an ultrasonic sensor).

Another fine feature of this versatile probot is its ability to carry payloads of up to ten pounds. B.O.B.'s maximum payload is half that of the Topo robot because of B.O.B.'s increased weight. Nevertheless, with the optional $95 Androwagon that may be purchased for either Topo or B.O.B., this probot can carry drinks, snacks, or even toys from room to room.

Besides its autonomy, another of B.O.B.'s attractions is its expandability. B.O.B. has a unique Androbus system that allows the probot to be expanded at any time. This guarantees that B.O.B. will not become obsolete for at least a few years, since add-on products may easily be installed with the Androbus system. There are already options available for B.O.B., such as the AndroFridge which can store cold drinks within the probot until you ask for them. Nolan Bushnell called the "beer program" his favorite, since it was very convenient for Nolan to ask the probot for a beer whenever he wanted one. An AndroSentry cartridge is also available for the probot, which will make B.O.B. roam about the house at night, safeguarding the home against burglars or other intruders. Finally, a voice recognition system is said to be in the works for B.O.B., which would allow you to give the probot voice commands.

One of the ten slots in B.O.B.'s Androbus can be occupied by the optional programmer package made by Androbot. This package allows a user of the probot to enhance B.O.B.'s onboard speech, navigating functions, and movement by means of a computer. The cartridge comes with the Androbot Control Language (ACL) which accepts simple commands, such as: MOVE FORWARD FIVE FEET. During the actual programming, the probot is connected to a terminal, but once the programming process is complete, the cable may be disconnected from the computer and B.O.B. will remember the commands given to it. In addition, B.O.B. also has a pat switch for receiving answers to "yes" or "no" questions. Therefore, if B.O.B. asked, "Would you like a drink?" a simple tap from its master would tell the probot "yes" or "no," and it could act accordingly.

As you can see, there are many exciting features and options available that make B.O.B. a versatile and useful probot. Still, it is more expensive than the HERO 1 and it doesn't have an arm. On the other hand, it is an easily expandable system with a great deal of promise for the future.

USES FOR BRAINS ON BOARD

Because B.O.B. has its own computer with a great deal of memory, there are many possibilities for this probot. The most important thing is to utilize all B.O.B.'s available power, including its sensors, voice, memory, and so on. With this in mind, a few potential uses for B.O.B. can be created, such as:

1. *Using B.O.B. as a Teacher.* With its voice and sensors, B.O.B. could teach just about anything. For example, it could teach young children the difference between "hot" or "cold" by using its heat sensor to detect different temperatures. If B.O.B. found something of a certain degree of hot or coldness, it could roll over to it and say with its own voice, "This is hot," or "This is cold." B.O.B. could also instruct children the concepts of nearness and farness with its ultrasonic range sensor. B.O.B. is also an excellent extension of a home computer, so both children and adults could learn more about their computer and B.O.B. by using the programmer package which interfaces a personal computer with the probot.
2. *Programming B.O.B. as a Party Host.* Besides the fact that a probot itself is a great novelty and quite a conversation piece, B.O.B. could be quite helpful at parties or gatherings. Using its heat sensor and ultrasonic range finder, it could be programmed to move to guests and ask if they would like a drink. B.O.B. could then either dispense a beverage from its AndroFridge or return to the kitchen to get a particular drink. An important aspect of a program like this would be for B.O.B. to wait for a short time after asking "Would you like a drink?" then leave if no response was received. This is an important factor for a program of this sort, since a furnace

or another source of heat (that B.O.B. thought was a person) would never respond to the question, and B.O.B. shouldn't have to wait forever to get an answer.

3. *Using B.O.B. as a Babysitter.* By using its sensors, the probot could make sure children don't leave a certain area. Children would also be more inclined to clean up after themselves if B.O.B. helped them out, perhaps by asking them to put their toys into his Androwagon so he could take the toys to the children's bedroom. Once there, B.O.B. could ask that the toys be removed. With this kind of unusual playmate, children might find cleaning up more fun that it ever was.

THE RB5X PROBOT

In August of 1982, the RB Robot Corporation of Golden, Colorado was incorporated for the purpose of designing, manufacturing, and marketing robots and robotics systems. Their first product, announced in September of 1982, was the RB5X Intelligent Robot (see Figure 5-2). It was the first mass-produced programmable robot made for home use, experimentation, and educational purposes. It is one of the few personal robots available that has components, accessories, and software already made for it. Some other ''probot companies'' are *planning* such additional accessories such as these for their probots, but unfortunately they haven't made them available yet. RB5X offers the most features, options, and the greatest versatility of any probot today.

MEET THE RB5X

The RB5X is said to be an ''intelligent'' probot, but this, of course, is not the case. Only a living brain can be truly intelligent, and it will probably be quite a while before an intelligent man-made machine is created, if ever. Nevertheless, the RB5X is intelligent in the respect that it has a tremendous amount of computing and mechanical power for a probot, which gives it the

Figure 5-2 Joseph Bosworth and his creation. *(Courtesy of RB Robot Corp.)*

versatility and expandability that most probot hobbyists would call for. The RB5X isn't going to perform any ''heavy duty'' functions such as doing the dishes, but it does have enough potential to make it rather useful and extremely educational and entertaining.

The features of the RB5X probot are mostly computer related, as you will see below. Its features include:

On-Board Microprocessor: Unlike some probots which require a separate computer to run them, the RB5X has its own ''microcomputer'' built into its system. The INS 8073 microprocessor makes the probot programmable and independent, but you do have to write RB5X programs using your personal computer, and then ''download'' them to the RB5X microprocessor for the probot's use.

Self-Learning Software: This small, first step toward true ''intelligence'' enables the probot to learn from its own mistakes. For example, you could set the RB5X down in a room and let it roam about randomly. It will probably run into walls several times, perhaps a desk, and maybe even a person. As it rolls around the room, it will ''learn'' in its own computer-like fashion where the obstacles are in a room, thus avoiding them in the future. The self-learning software are on ''Alpha'' and ''Beta'' levels, which were developed by the robotics author David Heiserman for the purpose of giving probots a simple way to ''learn'' from their experiences, somewhat like humans do.

Tiny BASIC: This subset of the BASIC language is the means by which a person can communicate with RB5X. Tiny BASIC is a subset of the high-level BASIC language, since many of the commands in BASIC aren't essential to operating the RB5X.

Sonar Sensor: The probot's Polaroid Rangefinder is able to detect objects in its path so the RB5X won't be running unnecessarily into any obstacles. The sonar sensor detection can be set from a 10-inch to a 35-foot range.

Tactile Sensors: The eight ''bumpers'' that encircle the RB5X allow the probot to detect when it has encountered an object. These sensors are essential to the self-learning software mentioned earlier, since the probot needs to ''feel'' when it has bumped into something so it can learn where objects are located. The tactile sensors also help RB5X navigate its way around its environment.

Battery Charging: Like the HERO 1, the RB5X can detect when it is running low on power. When it detects that condition, it uses its software to move about an environment back to its battery charger so it can "refuel" itself without your having to be concerned with it.

Software Module Socket: RB Robot Corp.'s probot also has an interface panel, which will allow you to plug in software modules with pre-made programs of 2K or 4K. The probot also has a built-in utility software module, which allows the RB5X to check its own battery power and standard software routines.

Dual RS-232 Interfaces: These two special ports allow the probot to communicate with other computers as well as other robots (or probots).

Remember, the probot does not have its own keyboard or keypad, so you must program the RB5X with your personal computer or be content with the ready-made software packages designed for the probot. Here are some other specifications concerning the RB5X in which you might be interested (see Figure 5-3).

Dimensions: 13 inches in diameter, 23 inches high

Weight: 24 pounds

Speed: Approximately 4 inches per second, using rubber drives and castors for motion

Computer: National Semiconductor INS 8073 with 8K random access memory (RAM) expandable to 16K or 18K

Programming Languages: Tiny BASIC, Robot Control Language (RCL), and optional Savvy language

In addition to all of these exciting features, there are a number of options available right now for the RB5X which you might want to purchase, should you acquire the machine. Some of these options may seem more attractive to you than others, but all of

Figure 5-3 The RB5X Intelligent Robot. *(Courtesy of RB Robot Corp.)*

them tremendously increase the probot's potential. Here are the various options, ranked in order of importance.

RM Arm: This robotic arm, which can be fully retracted inside the probot's body, can carry a load of up to 16 ounces and move it about in a wide variety of directions (see Figure 5-4). You can use a controller (somewhat like the "teaching pendant" of the HERO) to move the arm, or control it directly with software. The arm gives the RB5X the ability to carry and manipulate physical objects, which is essential to a practical probot.

Voice/Sound Synthesis: This option comes with a speaker, a printed circuit board, and a pitch/volume control so you can make the RB5X produce sounds as well as words. With this option, the probot can act as a burglar alarm, an educational device, or an entertainment machine.

Extended Memory: In order to put larger and more complex programs into the RB5X probot, putting 16K or 18K extra of RAM to its existing 8K will be necessary.

Voice Recognition: This option is available only for Apple owners. You can give the [illegible] such vocal commands as "turn [illegible]," "stop," or "get the newspaper."

RCL and Savvy: The two languages called Robot Control Language (RCL) and Savvy use simple English words and phrases to control the RB5X. This option is available only to owners of the Apple II+ and IBM PC computers.

Power Pack: RB5X already has a power pack of its own, but you can extend the "charge life" of the probot by purchasing this 10-ampere battery, which will keep the entire probot running for ten hours or the RB arm alone running for two hours.

SOME SAMPLE SOFTWARE

The software which is currently available for this probot is inexpensive, educational, and entertaining. Here's a sample:

1. *Pattern Programmer:* This module costs $19.95 and allows you to press combinations of the probot's tactile sensors to program a specific pattern for the probot to follow. This is an easy way to direct the probot around your home or busi-

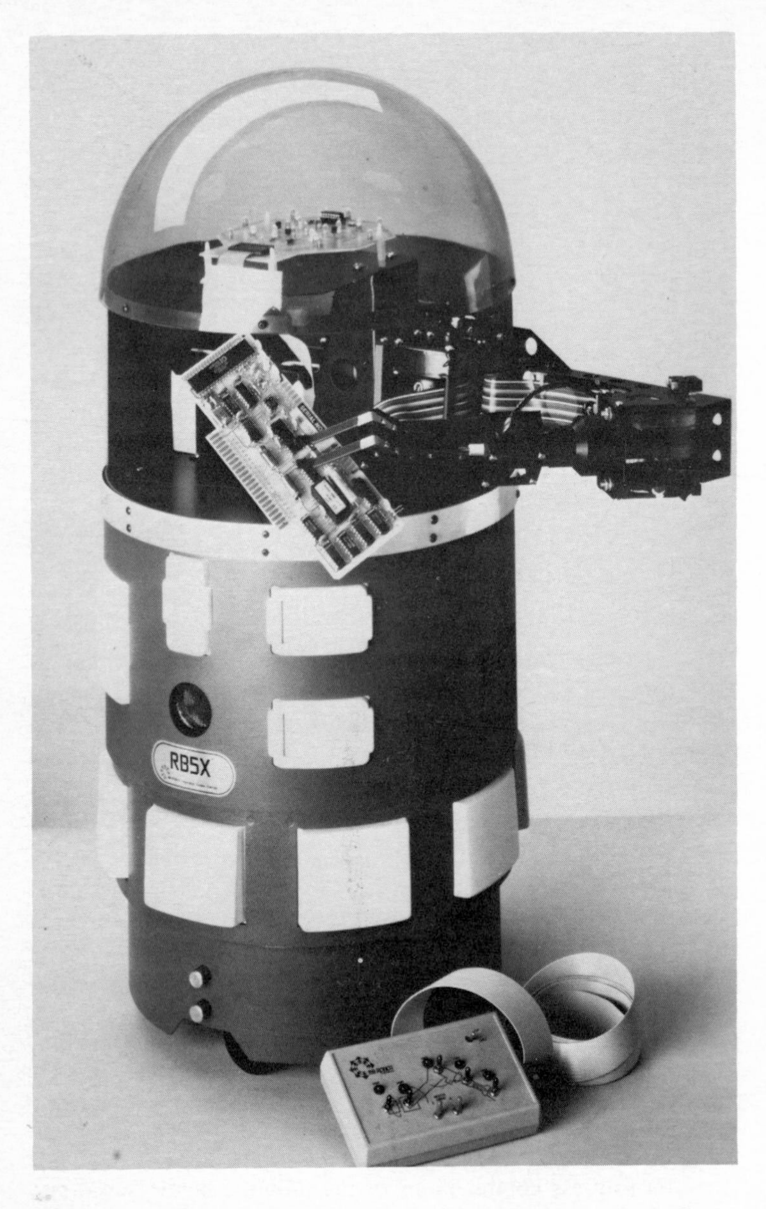

Figure 5-4 RB5X probot, showing the optional arm. *(Courtesy of RB Robot Corp.)*

ness, although you won't learn as much about computers as you would if you programmed the motions yourself.

2. *Bumper Music:* This isn't exactly the most useful program you'll ever find since it only plays a musical note when you press one of the bumpers of the RB5X. Each of the eight bumpers produces a different note, allowing you to create music by pressing the bumpers in different combinations. The suggested retail price is $14.95.
3. *Spin-the-Robot:* This game allows children to "interact," as RB Robot Corporation put it, with a probot equipped with voice/sound synthesis. It costs $14.95.
4. *Intruder Alarm and "Daisy, Daisy":* Probably the best module offered by RB, this software package contains two separate programs for the RB5X equipped with the voice/sound synthesis option. Intruder Alarm gives the probot the ability to sense movement within its sonar range. If any motion is detected, it sounds off an alarm. "Daisy, Daisy" is a program which makes the RB5X sing the song made famous by Hal 9000 in Stanley Kubrick's film *2001: A Space Odyssey*. If you have seen the movie, the song will evoke an eerie memory when RB5X sings it.
5. *RB5X Terrapin Logo Translator:* This is a software system that allows RB5X to execute turtle graphics procedures, making it an educational tool for demonstrating the physical capacities of the robot. Using the "turtle graphics" of LOGO, you can program RB5X to move exactly as the turtle moves across your computer screen. It costs $34.95.
6. *Recharge Cable Kit:* This allows RB5X to recharge its batteries via a 6-ft phone jack cord connected to its charger, which lets the user run programs that involve movement while the robot is charging. The retail price is $29.95.
7. *Hop to It:* At $24.95, this engaging educational game allows RB5X to use its sonar sensor to challenge players to judge distances in feet and inches.
8. *Math Whiz:* This is a math quiz that can be played by up to 8 people. RB5X uses its random number generator to compose a math problem for each player. It checks their answers for errors, corrects or congratulates, and calculates scores. It costs $24.95.

When RB Robot Corp. first introduced the RB5X probot they made a "sculpture" of six probots performing different tasks. One probot sang "Daisy, Daisy," another greeted visitors as they entered the display, one served as a "barker," trying to attract customers, another watered plants, and two others passed a bouquet of flowers back and forth. This small example of robots interacting together shows that software is already advanced enough to make robots entertaining, interesting, and useful, so the more advanced software of the future should be truly amazing.

THE FUTURE OF RB5X

If you are excited about what you already know about the RB5X probot there are some even more amazing features being developed. Here are some of the things which you can anticipate from the RB Corporation.:

1. *Fully Resident Voice Recognition:* Instead of requiring an Apple computer in order that the RB5X recognize your voice, the RB5X will soon have its own "built-in" voice recognition system. Because of this, you can issue commands directly to the RB5X without having to type in commands from a computer.
2. *Vacuum Cleaner Attachment:* Believe it or not, there is actually going to be an economical probot vacuum shortly. Unlike most $3,000 probot vacuums available at exclusive and expensive specialty stores, this attachment will make the RB5X a "self-contained vacuum" for an undisclosed (yet reportedly reasonable) amount of money.
3. *Radio Communications:* In case you want to program the direction the RB5X moves via a radio controller, this peripheral will be available in 1984 from RB Robot Corp.
4. *Trailer:* Like Androbot's Androwagon, this trailer will allow the RB5X to carry around light payloads.
5. *Navigation Detection:* This will be a supplement to the RB5X to make it more efficient in its movements about an area so it can learn more quickly how to avoid obstacles and reach a predetermined destination.

6. *Fire Detector and Extinguisher:* Like the R2-D2 robot in *Star Wars*, the RB5X will be able to detect a fire and put it out (to the best of its abilities) with this option.
7. *Mobility Package:* If you need to transport the RB5X and its accessories, this package will allow you to do so without possible damage to the equipment.
8. *Fully Resident Robot Control Language:* In the near future, RB Corporation's probot will have the RCL built into it so you needn't purchase the extra software for your computer.

Even without these future options, the RB5X is indeed, in my own opinion, the best personal probot available today for the money. It is extremely useful because of its optional arm (see Figure 5-5), voice synthesizer, and programmability, and it has enough options and future accessories that it should remain an important and popular probot in the personal robotics industry for several years to come. It seems that each probot has pluses and minuses, and to give you a brief look at the basic facts about each probot, all six will be re-examined at the end of this chapter.

ROBOTIC RESEARCH

We have taken a general look at both industrial robots and personal robots. We have examined the uses for these machines, and some of their potential for the future. However, the question "Where do these robots come from?" might be on your mind, so now we'll take a brief look at the work behind the creation of robots.

Many different kinds of technologies come together in the creation of a robot. Computers, hydraulics, electronic sensors, motors, and electrical systems are just a few of the pieces which make up a robot. Because the prices of high-technology products, such as computer microprocessors, have dropped drastically over the years, robots themselves have become less expensive. Research and development on nearly every component of robots, especially electronic and computer components, is becoming more active as the value of robots is becoming apparent.

Figure 5-5 RB5X delivering the morning paper. *(Courtesy of RB Robot Corp.)*

Because of the enormous complexity of a robot, however, we can't examine the design, manufacture, and installation of every component of a robot, either industrial or personal. There are too many robots and too many parts that make up a robot to be discussed here. Still, to get an idea of the research and development behind all robotics technology, we can examine one specific part of a robot: the arm.

The robotic arm has been, up to this point, the most useful and flexible part of a robot. Almost every industrial robot utilizes an arm, for welding, painting, or assembling, and the usefulness of any robot, even a probot, is greatly diminished if it has no arm.

Interestingly enough, we learn to appreciate our own bodies even more when we try to make machines to do the work of our bodies. Most people don't have a great deal of interest or awe about their own arms, but these limbs are incredible machines in themselves. Imagine trying to construct another arm, like your own. It would also have to move about and rotate just as your own arm does. If you think of the different ways your arm can move, its versatility becomes apparent. Here are some of the motions needed to simulate an arm:

1. Elevating and lowering the entire arm
2. Back and forth motion
3. Retraction and extension (reaching in and out)
4. Rotation of the arm (twisting)
5. Moving the upper arm up and down
6. Rotation of the arm at the shoulder (circular motion)
7. Rotation of the hand at the wrist

The list becomes even longer when we consider the hand, which is probably the most useful part of the body. The hand is versatile enough to draw, sculpt, type, grasp, shape, write, and manipulate. Because of the tremendous complexity of the human body, robotic arms have been difficult to perfect, let alone robotic hands.

However, quite a bit of progress has been made. Currently, there are three different types of robotic arms: hydraulic (using oil), pneumatic (using air), and electrical (using current). These first two are expensive, noisy, and bulky, so they are used mainly for factory work. Electrical arms are used in factories, too, but are also used for smaller mobile robots, such as probots, since they are quiet and relatively inexpensive.

A human arm has basically three joint locations: the shoulder, the elbow, and the wrist. Of course, the arm is capable of so much because of its inner complexity, consisting of muscles and bones. Most scientists have not tried to copy the human body because it's so highly sophisticated. Instead, different approaches have been taken to simulate certain parts such as the arm (see Figure 5-6).

Robotic arms, for example, tend to have quite a few axes of rotation. An industrial robot might have axes for horizontal travel, vertical travel, column rotation, elbow bend, extension/retraction, shoulder bend, wrist rotation, and wrist bend. Robotic arms tend to have numerous axes because the number of intricate motions that a robot's arm can perform is determined by the number of axes it has. Each axis is called a "degree of freedom," so a robotic arm with nine axes would have nine degrees of freedom. Another important reason for having so many axes is that an arm is not a whole body, so the robotic arm must have added versatility to compensate for that fact.

It is true that robotic arms today are capable of performing relatively intricate movements. However, to enhance the performance of the robotic arm, some researchers are working on inventions that will make the arm even more useful, such as artificial skin and stereo vision.

John Purbrick, of the Massachusetts Institute of Technology, has a particular interest in artificial skin for robots. This skin would have a sense of touch; it would be able to feel pressure being exerted. Artificial skin such as this has tremendous potential, since robotic arms would then be able to sense just how tightly they were grasping an object. For example, a delicate light bulb could break if grasped too firmly or a heavy piece of metal would fall out of the robot's grip if held too lightly.

The tactile sensor, or skin, on which Purbrick has been working is just one of the many ideas he keeps in his notebook. He also hopes that this skin will go beyond being just pressure-sensitive, and that it will eventually be able to identify objects

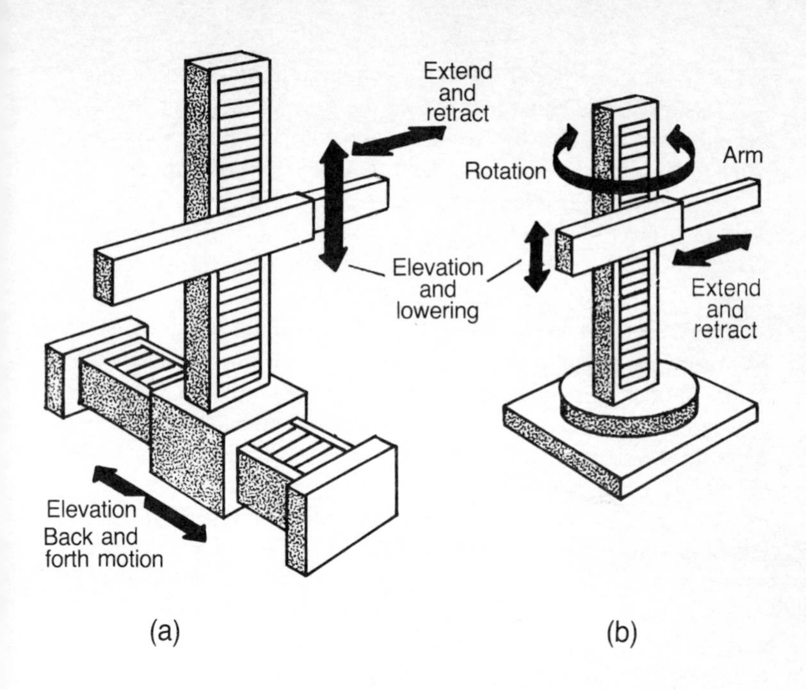

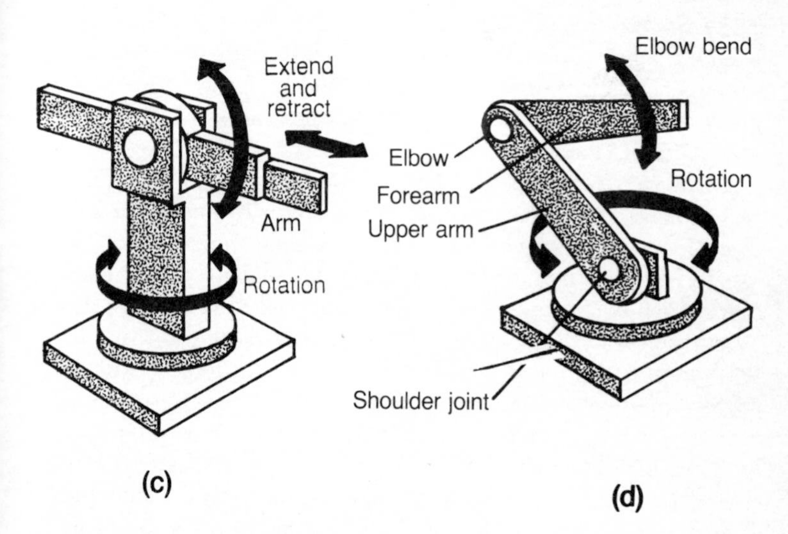

Figure 5-6 Four kinds of robotic arms and their motion.

solely by touch. A touch-sensitive robotic arm could know when it makes mistakes (such as missing the hole where a piece of an assembly was to be inserted) and make compensations to correct it. Most experts agree that the next wave of industrial robots will be dependent on artificial skin, especially in the automobile industry.

So far, however, there has not been a great deal of progress made in developing artificial skin. Although Purbrick has created a skin of sorts which can detect at least 10 different amounts of pressure in a range of 5 to 100 grams, it took him several years to accomplish this. His experience shows how difficult and time-consuming development in robotics can be. And, even with the work Purbrick invested in the tactile sensor, it still has some problems. For instance, the skin loses sensitivity over a period of time since it deforms with use. It is still a step, though, in the path to perfect a usable and practical artificial skin for robotic arms.

Yet another useful device being developed to enhance the performance of robotic arms is stereo vision. We perceive depth by combining the images from each of our eyes, since each one sees objects at a slightly different angle from the other eye. Because of this, we see three-dimensional images. Stereo vision would be a great enhancement to robots and the use of their arms, since perceiving texture and shape would give the robotic arm many more uses, such as being able to pick out specific objects at specific distances.

Seeing two images at once is no problem for a robot, since two cameras are all that is needed. The difficulty for the robot is determining if the object it sees with one camera and the object it sees with the other camera are the same. When you look at an object from two different angles, that object usually appears different each way (see Figure 5-7). View an object with first the left eye and then the right. Each eye sees the object from a slightly different angle. For the past twenty years, researchers have been trying to find a way to allow robots to jump this hurdle, but have had little success.

M.I.T. researchers, however, have found a way for the robot to determine if objects seen by each camera are the same. This method, using a "convolver" (a camera-like device with motors

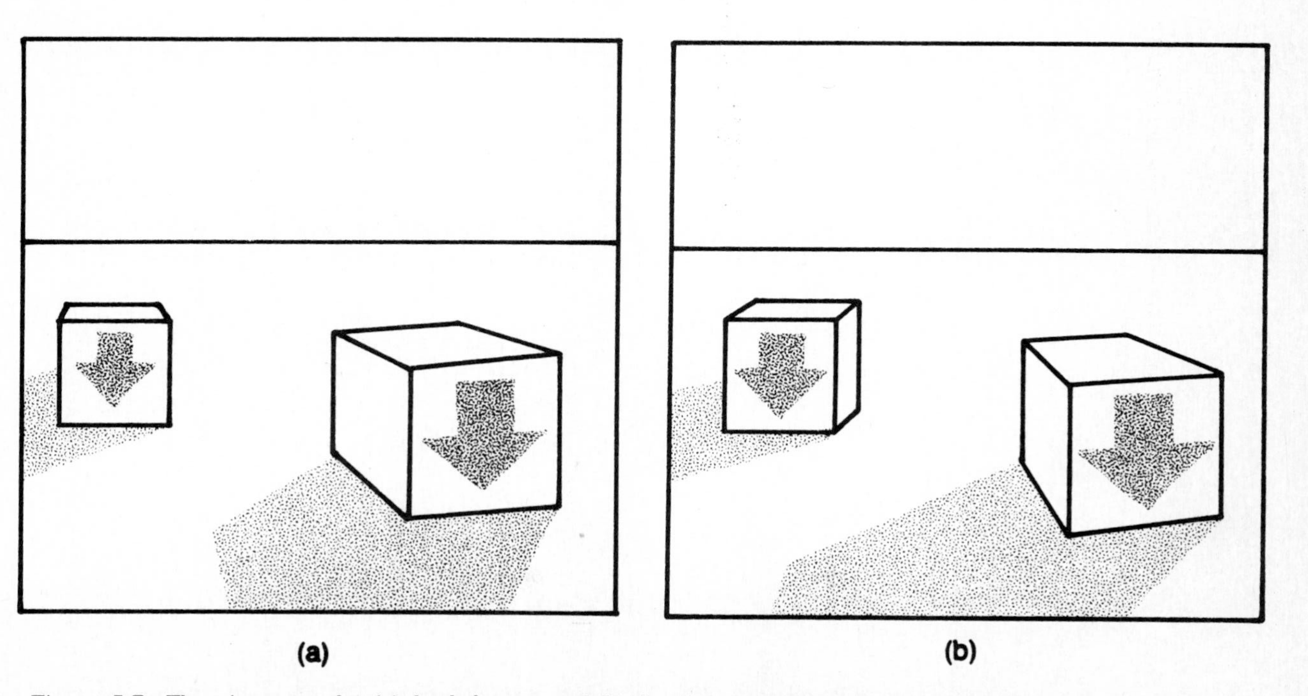

Figure 5-7 The view seen by (a) the left eye, and (b) the right eye. Stereo vision creates a perception of depth by fusing the two side-by-side views into one coherent three-dimensional picture.

to adjust individually the positions of its lenses), smooths away the fine detail of an image, marks the places where edges, bumps, and other important "depth features" exist, then compares images from each camera. If these rough images match each other, the robot can make specific calculations for that object's three-dimensional characteristics.

With the improvement of the robotic arm and the development of artificial skin and stereo vision, it seems that industrial robots and probots have bright futures ahead. Robots will become more useful with inventions such as these, and this brief look at some of the developments going on in robotics shows just a small portion of the work involved in creating a complete robot. Robotic research will gow, and more than likely flourish, but the people currently working in the field of robotics will probably have a great influence over this new technology for many years to come.

THE PROBOTS IN REVIEW

The six probots we have examined in the last three chapters have varied in price, performance, and potential. For those interested in eventually purchasing a personal robot, however, whether for business, home, or school, a brief review of each of the probots might be helpful. Therefore, here are some facts about each of these machines:

AndroMan: This mini-robot from Androbot, Inc., is designed to play games through the Atari VCS 2600 system. It comes with a game cartridge and three-dimensional landscape, and soon a whole line of video game cartridges will be available for it. AndroMan also has a forty-five word vocabulary that he uses while playing this video screen/real-life action game. Approximate price: $199.

F.R.E.D.: The Friendly Robotic Educational Device from Androbot is a foot-high machine, controlled by a infrared signal and a personal computer. It is equipped with sensors to make sure it doesn't fall off a desk or down the stairs while it moves about, even when carrying a payload. F.R.E.D. also has a forty-five word vocabulary, and it can hold a pen and draw just about any shape or design you program into your home computer. As with most of the personal robots from Androbot, Inc., software packages are being made available for this small but amusing probot. Approximate price: $299.

Topo: A personal computer is needed to use Topo, the three-foot high Androbot that uses TopoBASIC, TopoFORTH, or TopoLOGO. Topo can talk, carry up to twenty-three pounds, move about a home (remembering the layout of the building), and help you learn more about robotics and computers. Topo's voice is an option, costing about $200 extra, but it is well worth the price, considering the additional uses the voice makes possible. Approximate price: $795 (without voice).

HERO 1: Heath's Educational Robot, one of the first to appear on the personal robot market, has nearly everything a probot owner could want: a voice, direct programmability, an arm, sensors, and mobility. This probot may be purchased in either kit form or assembled, but a kit will take aobut eighty hours to put together. Although this robot certainly isn't as "cute" as some other probots, it is extremely functional and well-documented. Approximate price: $2,495 assembled, $1,495 unassembled.

B.O.B.: Androbot's Brains On Board machine is exciting, expandable, but relatively expensive. However, considering the features of this probot, it is well worth the price. Sensors, autonomy, three megabytes of memory, two Intel 8086 microprocessors, and sophisticated mobility make this one of the "smartest" probots that have been created. It is also a very expandable machine, with ten ports in its Androbus system, which can be fitted with options, such as the programmer package. Although B.O.B., like its Androbot brothers, lacks an arm, its maker promises that an arm will be available soon.

RB5X: This probot, which looks a lot like the R2-D2 robot from *Star Wars*, has an optional arm, voice/sound synthesizer, and expanded memory. Its standard features include a built-in microprocessor, eight "bumpers" that detect when the probot touches something, and a motion sensor. A number of exciting add-ons are going to be available soon for this machine, like a fire detector/extinguisher, more software modules, and a burglar alarm program. This is probably the best buy among all the probots mentioned in this book. Approximate price: $1,200.

Having examined all of these probots, as well as some of the development going on in robotics, we can begin to explore some of the practical uses and long-term impact of personal robots on our society, our economy, and ourselves.

CHAPTER SIX
THE PRACTICAL PROBOT

The fact that personal robots are machines of sophisticated technology is undeniable. The robots of today, though few in number, possess exciting and useful features and functions that will be improved and expanded upon even more. Still, when some people are first told about probots, their question is a skeptical "So what?" As mentioned in Chapter 1, these skeptics will be fighting probots for some time to come; they are convinced that because we have survived without personal robots for this long there is no need for them now.

Indeed, we could go on living without probots. For that matter, we could survive without telephones, television, cars, refrigerators, airplanes, and the other things considered necessities in our society. The point is that luxury eventually becomes necessity. For example, humans got along without telephones for thousands of years, but what would happen if all the telephones in the world suddenly disappeared? How could you contact someone in an emergency? What if you wanted to talk with your friends? How could businesses survive? Obviously, telephones are an essential element of everyday life in our society.

If we could ask a person living in the year 2084 what the world would be like without robots, he might shudder and ask, "What would happen to industry? Our economy is dependent on the robot! Who would do the housekeeping? Who would teach the children? Who would make the cars, build the houses, maintain the landscape? And what would happen to the National Robotic Football League?" Even though our society is doing well without robots at this time, one day the thought of living without them could seem frightening.

The point of this chapter, however, is not merely to convince skeptics of the practical uses for the probot; it is also an attempt to suggest functions which you might wish to program into your personal robot, should you acquire one. You may even want to create a program which performs a particular function, then sell it to probot owners. Finally, this section of the book is intended to stir up some of your own creativity so that you might think of some original uses for probots.

UTILIZING THE PROBOT'S POWER

To take advantage of any probot's full potential, you must understand and use its power fully. If the probot has an arm, the arm should be used to its greatest potential. If a probot has a variety of sensors, each one should be used as much as possible. The reason for this utilization of power is to get the most for what you pay and to buy only what you need.

Like personal computers, personal robots vary in price and features. A small probot like F.R.E.D. is inexpensive, but it doesn't have very many functions. On the other hand, the HERO 1 has a wide range of functions, so that programs made for it can be more varied and useful.

Whatever the features on your probot, here are some suggestions for using some of the most popular equipment supplied with the more expensive machines:

Sensors: There are several different kinds of sensors, and each one can serve several different purposes. A *heat sensor,* for example, can be used to locate people (since people tend to be warmer than their surroundings). It could warn you of a stove that was left on, or even wake up the occupants of a house if a fire breaks out.

An *ultrasonic range finder* can be used to determine specific distances. The probot could find the distance between itself and another object or a person, and go exactly that far to reach the object. A range finder could also help a probot guide itself around a house.

A *light sensor,* used to detect the presence and volume of light at any given time, could let a probot know if lights were on or off. This might be useful for a program in which the probot roams the house while its owner is on vacation, turning lights on and off to give the appearance of a person in the house, thus frightening away potential burglars.

A *sound sensor* can tell the probot how loud a sound is so it can take appropriate action. On a Saturday morning, for example, when parents might be trying to sleep, a probot could tell children "That TV is too loud." A sensor of this sort could also detect peculiar noises, such as a burglar trying to break into a house. If the probot did hear a sound, it could remember the time when the sound was made and go to the source of the noise to investigate.

Lastly, a *motion detector* can also be an excellent protection device, since the probot could sense any movements being made within a certain area, such as the motion of an intruder late at night. Or, the probot could make sure a person was awake and out of bed on weekday mornings by determining if motions were being made to get out of bed and off to work. If no motion were detected, the probot could take appropriate action, such as saying, "GET UP OR YOU'LL BE LATE FOR WORK! IT'S NEARLY HALF PAST EIGHT!"

Mobility: The feature that separates a probot from a computer is mobility. The probot can move in a circle, forward, backward, or in whatever direction you specify. The mobility can be used to search the house for intruders, carry beverages to guests, or move toward a noisy TV set to turn the volume down.

Arm: Although arms are not standard features on many of today's probots, they have a great number of uses to offer. Arms can pick up and put down objects, draw, assemble parts of an object, or shake hands with a human. Arms can move in practically any direction, and when coupled with other features of a probot, the arm can be a very useful tool.

Voice: Today's synthesized voices don't sound very human, but they add many practical uses to a probot. The voice is an efficient and practical communications link, understandable by practically everyone. A voice can be used for education, notifying the occupants of a house about an intruder, or making interesting sound effects.

Computer/Memory: At the heart of every probot is a computer with memory. Sometimes the probot has a computer of its own, as HERO 1 and B.O.B. do, while others have to be connected in some way to a personal computer. The memory can be used to store information, such as foreign words, multiple programs, or subroutines to be used by the computer. Like a probot, the computer should also be used to its fullest potential. If, for example, the F.R.E.D. is used with an Apple II, the Apple should be programmed to draw elaborate shapes on the screen as well as make F.R.E.D. draw patterns on a piece of paper, since that computer has the capability to draw high-resolution graphics.

Once you understand how to use these features to the fullest, you can make useful and educational programs for your probot.

USES FOR A PROBOT

The uses for a probot are limited only by its features and your imagination. We have explored some of the features, so it's time for your creativity to do some of the work. To give you a few ideas about how a probot might be used, both today and in the near future, some suggestions follow that you might want to consider.

EDUCATION

Robots, like computers, are excellent teaching tools. Because of a probot's memory, voice, and ability to accept information from a human, it can teach practically any subject. For instance, a probot could be programmed to read stories to a child. The stories could be in the probot's memory, and the child could request a favorite tale. If coupled to a computer terminal, the probot could display the words to the story as it is read so the child could read along.

Or, foreign language could be taught to a human, by using the probot's voice. The probot could recite particular words, first in English then in the foreign language. The keypad or keyboard to the probot could be utilized as well, perhaps for having quizzes on how to spell certain foreign words or phrases. It could even serve as a translator, converting the English typed into the keyboard into another language, spoken with its voice synthesizer.

One of the best subjects a probot can teach a person is robotics. A person using a probot learns how to work a robot as well as a computer. Knowing how to use these high-technology devices will be an asset as our society grows more dependent on computers and robots.

HOME USE

The personal robot is meant for the home, therefore, there are quite a few possibilities in that type of environment. Since a probot can carry payloads, you can program it to carry just about any object, as long as it isn't too heavy. It could carry drinks to party guests, move toys from one room of the house into a child's bedroom (Figure 6-1), or even serve as a party host, mingling with guests and serving snacks occasionally.

At dinner time each day, the probot could roam about the house, search for people with its heat sensor, and announce, "Dinner is ready!" to anyone it finds. Also, with the probot's mobility and memory for predetermined paths, it could even take the dog for a walk. This wouldn't be a great idea in a city, of course, since someone could just steal the probot as well as the dog. Risks such as these must always be considered before programming a probot for a particular function.

Although probots do well at repetitive, menial tasks, they can also excel in other areas of work, even babysitting. A probot, properly programmed, could keep watch over several children at one time, making sure they don't leave a specified area. It could entertain the children with a game of hide-and-seek, or "watch" television with them, and put them to bed at a specified time. Of course, with present robot technology, you probably would not want to leave a child attended only by a probot. But eventually, probots will be sophisticated enough to care for humans without any supervision. The greatest advantage to a probot babysitter is that you won't come home to an empty refrigerator and a bill for services rendered.

SENTRY AND PROTECTOR

A probot is very loyal to its owner, and when programmed properly it can guard its master and its master's possessions. Using its mobility, its memory for knowing paths around a house, and its sensors, a probot can roam around the house at any time, listening for noise and watching for motion. If something unusual is spotted, the probot might cry out in its robotic voice, "PLEASE IDENTIFY YOURSELF!" If you were just getting a late night snack, a response such as three quick handclaps could let the probot return to its sentry duty. However, if the proper response was not given, the probot could cry out in its loudest voice, "INTRUDER ALERT! INTRUDER ALERT! GET OUT OR

Figure 6-1 Topo and Androwagon. *(Courtesy of Androbot, Inc.)*

I'LL BLAST YOU WITH MY PHASER!'' followed by some noises that sound like a laser gun. The burglar will very likely run out of the house in terror (Figure 6-2).

Using its heat sensor, a probot could also make an excellent fire alarm. If excessive heat were detected, it could go from room to room making noise and shouting, ''FIRE!!!'' You could preprogram the probot to ignore the areas of your home, where radiators might be, since excess heat would be detected in those spots.

Finally, if you left your home for a vacation, a probot could be programmed to give the impression that someone was still at the house. It could turn lights on and off infrequently, listen to a radio for a while, turn on the porch lights at a certain time, and so on. In addition, the probot could still perform its sentry duties in case a burglar breaks into the house.

ENTERTAINMENT

Some people think of probots as rather boring, monotonous machines. Actually, these high-technology devices can be very entertaining. Some probots, like AndroMan, are made exclusively for entertainment. Others, though made for many different tasks, can still provide a lot of fun for those who use them.

F.R.E.D., for instance, can draw just about any kind of pattern onto a sheet of paper. Other probots can play games like tic-tac-toe or chess, using their internal computers, and with some games, their arms. Nearly any game it's possible to play on a computer (with the exception of video games) can be implemented on a probot, with the added advantage of the ability to move objects, such as playing pieces on a board.

In the future we may see even more advanced games played by probots, such as basketball and soccer. Perhaps we could even program the probots to play against themselves, to see which programmer devised the better strategy. Another exciting possibility is that the strategies of famous players could be programmed into the robots. You could, in fact, play a game with the equivalent of a champion player. Of course, you might not want to fight a robot programmed to box like Muhammed Ali or play soccer like Pelé, but it might be fun to try.

ALERT! IDENTIFY YOURSELF IMMEDIATELY OR I'LL SHOOT!
$

BUSINESS

Probots don't have to stay in the environment of a home to be useful. Think of what an interesting advertising machine a probot could be. It could roam about a store, telling people about a particular product that is new or on sale. In a computer store, a probot could point out the computers' features and power. Or, it could help answer questions about a company's products with preprogrammed answers.

The technology we possess today does not permit probots to be particularly helpful for things like filing, making calls, or coming up with a new sales strategy, but eventually computers and robotics will become advanced enough for some work of this type. Right now, a probot is an interesting advertising aid to use in business, and an excellent way to attract attention.

You may have thought of some other uses for a probot while reading these suggestions, or you might even have had some ideas before reaching this chapter. Just think about things that you would rather have someone else do for you, and consider whether a probot could do them. Do you hate doing your taxes? A probot can't help much there, but a computer can. Do you dislike vacuuming? Well, that's one thing some probots can do. Keep in mind that the tasks you set for a personal robot must be within its capabilities, and as time goes on, those technological boundaries will grow wider and become less restrictive on your ideas.

THE POTENTIAL FOR PROGRAMMERS

The computer industry has spawned many other businesses in its wake. Companies that manufacture peripherals such as printers and disk drives, consultants who select the computer best suited for a person's needs, and firms that write programs for computers, are all doing well. The reason for their success is the great demand for computer products. Since computers are selling well, computer-related products are, too.

In the same way, robotic products will be selling well as the probot revolution continues. One particular product that will be greatly in demand is software for probots. Anyone who can write software for a popular, useful robot should make quite a bit of money.

If you decide to write a piece of software for a probot (or robot), remember that the industry is still very young. Also, be sure to write for one that is popular if making money is your goal. In the computer industry, programs written for the best-selling machines, such as the Commodore 64, make their authors a great deal of money. If those authors chose less popular computers, they would have much less to show for their efforts. As far as probots are concerned, it is hard to tell which ones are "hot" and which are going to eventually fade from the market. The secret to successful software, though, is to write a useful program for a product that is popular, has a lot of features, and is expandable enough to *remain* popular.

There are few companies manufacturing probot software, but they will be increasing in number shortly. The demand for quality software will attract entrepreneurs to this new industry, and if you would like to be a part of this potentially enormous business, you should begin learning as much about your probot as soon as possible. With the knowledge you gain, you can write high-quality, practical programs which you can sell to those who need them.

IMAGINATION AND PROBOTS

The key element to making probots useful is imagination. It may seem strange having to think of ways to make a machine useful. But, as noted before, many new inventions seem unnecessary at first. When they are used enough they become "essential." When personal computers first appeared, many owners had trouble answering the question, "What can it do?" Today, there are thousands of uses for computers, and there will certainly be many more to come.

After all of these ideas about using probots have been suggested, some skeptics may still wonder if personal robots are really necessary. All I can say to them is, "You'll see!"

CHAPTER SEVEN
PROBOTS FOR TOMORROW

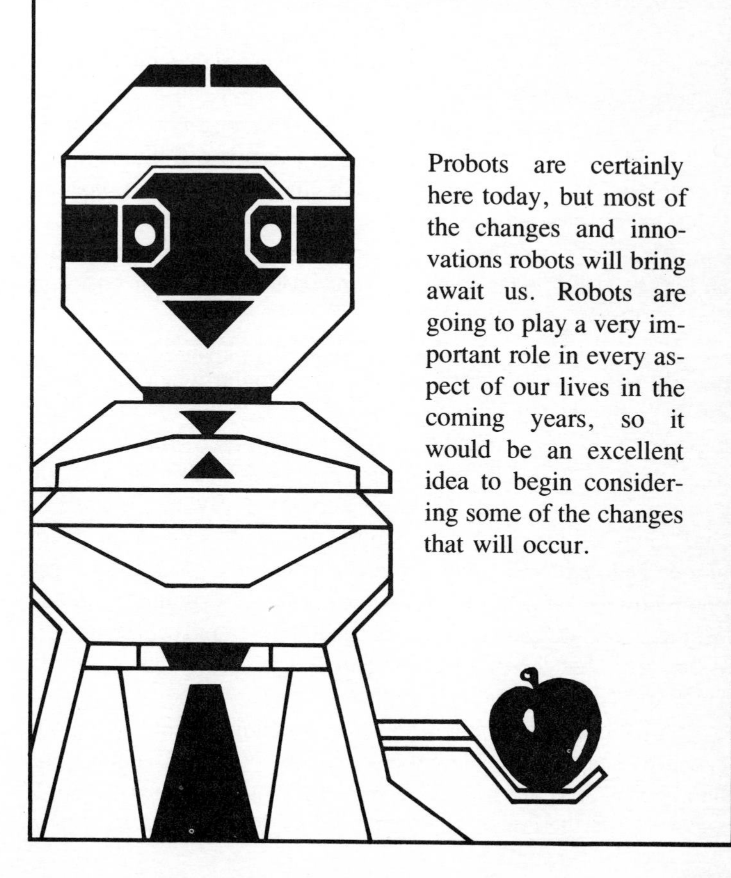

Probots are certainly here today, but most of the changes and innovations robots will bring await us. Robots are going to play a very important role in every aspect of our lives in the coming years, so it would be an excellent idea to begin considering some of the changes that will occur.

THE FUTURE OF THE ROBOT

Before we begin hypothesizing about our future, we should consider what the future of the robot will be like—that is, how the robot itself is going to change over the next decade.

Probots, like home computers, will drop in price as they grow more sophisticated. There are many reasons for high-technology products to drop in price, including the lower cost of memory and microprocessor chips, increased competition in the marketplace, and expanded, more efficient production methods.

When pocket calculators came on the market, around 1973, the most basic one cost over $100. Ten years later, a much better calculator could be purchased for less than one-tenth of that price. Following the same pattern, computers made in 1975 cost several thousand dollars, yet today computers of far higher quality may be bought for $300.

It is predicated that probots and industrial robots, like their high-technology counterparts, will drop in price in the next few years. One estimate suggests that industrial robots costing $150,000 in 1980 dropped in price to $100,000 in 1983, and will continue to decrease in price, costing $35,000 in 1985 and only $5,000 by 1990. The average robots could very well cost less than $500 by 1990 also.

Since the costs of robots should fall, sales should consequently rise. Just as computers became popular when their price fell to a certain level, robots should sell in huge numbers once their price becomes affordable for many people. The population of probots will reach a number even greater than that of the personal computer. When the home computer revolution was just beginning, it was said that computers would appear in every home. That may still happen, but computers are not seen in every home now because not everybody needs one. Computers can do wonders for investors, business managers, and video-hungry youngsters, but there are many people who really don't require the help of a computer.

A probot, on the other hand, is something nearly everyone can use, since it can manipulate objects as well as manipulate numbers. There are a very few people who couldn't use some extra help in the house, or a good babysitter, or a companion to play a game of checkers with, so probots are potentially more popular than the home computer. And a probot could still provide the capabilities of a computer, along with the ability to perform the physical tasks of a robot.

Besides the decrease in price and increase in number, robots will also grow in sophistication. As memory chips become smaller and microprocessors become more advanced, the abilities of the probot will grow in number and usefulness. The first computers didn't have very much to offer, despite their high prices. Today, however, home computers less than $300 have high-resolution graphics, color, multiple-voice music, sound effects, and a substantial amount of memory. Probots will probably fit this pattern, with more features appearing over the next few years, such as:

Voice Recognition: The ability to detect words or phrases on which the probot may act. For instance, if you told your probot equipped with voice recognition "Get the newspaper," "Serve the drinks," or practically any other phrase, the probot could recognize it and follow your command.

Advanced Robotic Arms: Today the HERO 1 and the RB5X are just about the only probots with an arm, but other, more advanced arms will become standard on most personal robots within the next decade. These arms, capable of more intricate movements and perhaps capable of "feeling" an object (with artificial skin) could make probots far more useful, since arms are probably their most vital working part.

Shape Recognition: When probots can recognize shape, they will be able to pick out a single object from a group or identify any object. For example, if a bowl were filled with many different kinds of fruit, you could tell the probot, "Get me an orange," and, using its shape-recognizing "eyes," it could travel to the fruit bowl and pick out an orange from the rest of the fruit. Of course, there are more applications for shape recognition, such as assembling pieces of an object or picking out a particular person among many by identifying the height, size, and build of an individual.

These and many other new features will be introduced in robots and probots as they continue to evolve. We have seen the many uses for the robots of today, so the number of uses for those of tomorrow seems almost too incredible to believe.

NEW INDUSTRIES

Human workers often feel threatened by robots, but actually they'll create more jobs than they destroy. And, more importantly, these jobs will stimulate creativity and enthusiasm within people, while the jobs of the robots will remain boring and monotonous.

Many of the new jobs will be in the field of robotics. Many of these new robot-operated jobs were discussed in Chapter 2, so I would now like to suggest some of the possibilities that entrepreneurs, rather than established companies, could exploit.

Robots themselves will be very much in demand, so it is likely that a large number of people will try making their own probot. A few will succeed, the great majority will fail. Enterprising people in the early days of the home computer, such as Steve Wozniak and Steve Jobs of Apple Computer, are extraordinarily wealthy today because they invested their time and talents in something they believed would succeed. Even giant Hewlett Packard was started in a garage (where else?) on the Stanford University campus about thirty years ago, and has since made its founders incredibly wealthy. There are many success stories for entrepreneurs; for those who create quality probots and market them successfully, the rewards should be great indeed. This will be especially true when probot sales begin accelerating due to lower prices and higher quality.

Robots cannot run by themselves, though, so robot software will also be a product in great demand. There will probably be quite a few companies making probot software for the more popular machines. These companies will discover the greatest uses for probots. Programs for home, education, and game playing will be especially popular at first, and the companies which enter the market soonest with the best products will survive the turmoil of this technological revolution, in which many companies are born and then die quickly.

Finally, products to support the probots, known as peripherals, will be needed. Home computers can use modems, printers, and disk drives, so many companies were formed to produce and sell these products. Probots will need extras, too, such as sensors, extra arms, more powerful motors, and other supplies which will enhance the machine's performance. Persons who repair and service probots, both industrial and personal, will also be very much in demand.

Some of the entrepreneurs forming these new robot companies will become rich, and perhaps even famous. Also, these companies will employ many other workers—including some who were displaced from their old jobs by robots. In this manner, the balance of work will be kept stable until such time as robots can even manage themselves—which is at least twenty years away. At that point, people will have to confront the new structure of their society, realizing that robots will be responsible for a large part of the work, and humans can finally enjoy their leisure time.

THE EFFECTS OF PROBOTS AND ROBOTS

How will the industrial robots and personal robots affect our society, our economy, and ourselves? Although many people are afraid of what robots might do to our way of life, as long as we control the robots, there is little to worry about. The trouble arises when we let technology get ahead of our wisdom, thereby letting the robots have more control over us than we have over them. Considering the great amount of political pressure that will be put on the makers of robots, however, it is extremely unlikely that robots will develop too quickly for our society to handle.

Most of the effects of the robotic revolution will be positive, despite some bad side effects. Some suggestions follow about what might happen to different aspects of our lives, in the age of the robot.

SOCIETY

Probots will substantially affect the way we live. They will help us shop in stores by picking out the products we specify. In hotels, they will serve as bellboys, maids, and clerks, making travel more efficient and less expensive. In fact, we could simulate travel by using Travel robots that could be placed in many parts of the world. A person in one country could sit in a specially designed booth, watching a screen. He could tell the machine that he wanted to visit another country such as Italy. At that point, a probot in Rome could step out of its holder, moving up and down the streets of the city as the viewer specified. The human could control the motions of the probot, roaming up and down any street in Rome, seeing everything through the probot's eyes. This may not be quite as much fun as real travel, but it's a fairly good simulation.

Society may take a while to adjust to robots, but these machines will eventually become as commonplace as the computer. Therefore, the sight of probots roaming about will become less surprising to people. Probots will not dehumanize us, as some people fear, but will be thought of in much the same way that computers are—as machines designed to make our lives easier and more productive.

EDUCATION

Teachers won't be replaced by probots (not for a while at least), but education will be supplemented by these machines. Children and adults alike will learn about robots, computers, and just about every subject through the use and programming of probots. Computers certainly haven't made human instructors obsolete, but they have been introduced into the schools as teaching aids. Once probots reach prices that schools can afford easily, it is likely that they will be even more welcome in the school environment than computers, since probots have so much more to offer.

HOME

Living in a house will be easier, thanks to probot sentries guarding against intruders day and night. Many of the household chores formerly done by humans will be handled by probots, such as cleaning dishes, vacuuming, and even washing windows. Although the home computer will still serve as the focal point for information, the probot will be the focal point for service, providing whatever labor is required within the home. If the dog has to be taken for a walk, the probot could do that. If members of the family must be awakened at different times of the morning, a family probot could do that as well.

Future probots will also be easier to program. For example, instead of having to code with hexadecimal commands through a keypad, a simple voice command will be all that is required. If, for instance, you wanted the probot to introduce your children to a new language, take out the garbage, and sweep the porch, you would only have to call out the commands to the probot, probably in a specific format. A dialogue between you and the robot (we'll call him George) for these tasks might sound like this:

Human: George!
George: At your service.
Human: George, follow these commands in the following order.
George: Ready to receive commands.
Human: Introduce the children to a new language.
George: Please specify language.
Human: Latin.
George: Task One acknowledged.
Human: Take out the garbage.
George: Task Two acknowledged.
Human: Sweep the porch.
George: Please specify which porch.
Human: Both.
George: Task Three acknowledged.
Human: End.
George: Will that be all?
Human: Yes, carry out your tasks.
George: Acknowledged.

It will be a while before voice recognition, robot technology, and software will be this advanced, but it is nice to imagine what it might be like to have an obedient, mechanical servant of this sort.

THE INDIVIDUAL

The psychological effects that robots will have on humans are difficult to predict, though some assumptions can be made how people will react to this age of the personal robot.

As mentioned previously, some people will fear the probots, as some did the computer. However, the fear of robots will probably be even greater, since the robot seems almost like another person due to its form. People usually fear computers because they think they will harm the machines, or the machines will harm them. People will fear robots for the same reasons, but to an even greater extent, since robots, because of their physical abilities, *can* actually harm a person. In fact, a woman recently won $10 million in a court case involving the death of her husband, who was struck on the head by a robotic arm. Safety standards for robots will certainly be rigid, but people may still fear the machines, suspecting that something might go wrong with their internal programming or mechanics.

Other people might even think of the robot as dehumanizing. Nothing could be farther than the truth, though, since robots lift humanity up from degrading work.

LABOR

At first, unemployment won't be a great problem, but job transition will. Many people will find themselves switching occupations more frequently, and the general trend for the future will be a job change for every person each decade. The nature of jobs will change as well. Robots will do the menial tasks that were formerly done by (unenthusiastic) human workers. These dangerous, dirty, and boring jobs will be handled almost completely by robots, while the humans will be transferred to more enjoyable work, better suited for the intelligent human being. Work will be generated by robots, since new companies will need labor of all types in order to survive. In fact, the problem will be a lack of labor, not a lack of jobs, so unemployment will certainly not be a problem in the first stages of the robot revolution.

When robots become more advanced they will be able to care for themselves. They will be able to manufacture and do many of the repairs on other robots, and a master computer will manage the robots. These machines will be advanced enough to make a great deal of human labor unnecessary. At that point, unemployment could become serious, so our society will have to adjust to the change. Shorter work weeks, more leisure time, and more creative, intellectual work will become facts of life in this new society of robots. Even though there will be some difficult times during this transition, the long-term effects of robots on labor will be more productivity, more leisure time for people, and a higher standard of living for the countries that utilize robots in the home and the factory.

There are those who will welcome robots with open arms. People who think of the coming of robots as a second industrial revolution will probably tell others, "To know them (robots) is to love them," since understanding is the key to appreciating any new technology. Generally, those who dislike robots will learn about them and learn to live with them, and within a decade or two nearly everyone will have accepted robots as powerful, helpful machines.

NEGATIVE SIDE EFFECTS

CRIME

All of our problems will not be solved by robots. In fact, these new machines will create some problems themselves. Or, to be more accurate, humans with bad intentions will make the robots cause problems.

There are many examples of good used for evil in the world of technology. Even the computer, a machine which has proved itself to be an amazingly helpful tool, has been used by individuals for bad purposes, such as embezzling from banks, breaking into other computers, and stealing information. Probots will probably also be used by some individuals for wrongdoing. If a probot can be used as a sentry, why can't a probot be used as a burglar? A bank robber? A hijacker? Criminals would be much safer controlling a probot rather than being at the scene of a crime themselves.

Criminals will probably come up with many ways to use probots for evil purposes. It's not the robots people should be frightened of, however, but the people who use them. The only way to make robots safe would be to make it impossible for them to harm anyone, or even threaten anyone, perhaps by using Isaac Asimov's three laws of robotics. However, the fact that robots could be used for bad intentions must not prevent their development. If every machine that could be used to do wrong were eliminated, what would we have left? There is no sound reason to halt the production of robots.

LAZINESS

Another potentially long-term negative effect of robots is laziness. As noted earlier in this chapter, a utopian robotic society would consist of robots doing the work, and humans spending their leisure time in creative, intellectual activities, or perhaps exploring the Earth or outer space. Still, not everyone in the world wants leisure time, since they wouldn't know what to do with it. Some individuals simply don't want to participate in more creative, intellectual activities, and if robots took over their jobs, they would become restless, bored, perhaps even criminal. There is no real solution to this future problem, except perhaps setting up artificial work, which *could* be done by robots, but is instead performed by humans. That is, those who still want manual work may have it.

THE BRIGHT FUTURE

SPACE EXPLORATION

Robots will not only help us on this world, but will also help us explore other worlds. In fact, space exploration is one of the most promising uses for robots, since these machines are not so susceptible to the elements that could hurt or kill people.

Robots can withstand radiation, heat, cold, and can function without an atmosphere. Also, robots have a much longer life span, so voyages to other solar systems, as well as other planets, would be possible. Robots could communicate with us, and if something went wrong and destroyed the space vehicle, a machine would be lost, not a human being.

SEA EXPLORATION

Mining the sea for minerals, lost treasures, and food are all within the capabilities of robots. As our robotic technology grows more advanced, we can explore the bottom of any sea, examining the mysterious creatures living in the freezing darkness, miles below the surface of the water. In general, robots could allow us to explore places where no person can survive. These machines of today and tomorrow can take us where we have never been before, and use untapped resources, previously unattainable by us because of our own physical limitations.

SOPHISTICATED TECHNOLOGY

A little closer to home, robots will provide all of the conveniences mentioned earlier in this book, as well as improved technology. In other words, the development of robots will lead to the development of other inventions as well, such as more advanced sensors and superior motors. The robot will not only generate leisure time, greater productivity, and lower prices, but will also lead to more sophisticated technology in other fields of science and industry.

PROBOTS, TODAY AND TOMORROW

There is little doubt that probots, robots, and "high-tech," in general, could provide us with better living than we have ever known. Robots, though introduced recently, are already making an impact on our society and our economy. These effects will grow along with the number of robots, and, if we manage our robots wisely, these machines of the present and the future can make all of our lives better (Fig. 7-1). We must remember, however, that we hold the power of these machines and the future of our society in our own human hands. Let us hope that we have the wisdom to keep it that way.

Figure 7-1 The probot is the key to a bright future.

BIBLIOGRAPHY

Albus, James S. "Robots in the Worldplace." *The Futurist,* February 1983, 22–27.

Androbot, Inc. "Topo Owners Manual." San Jose, California, 1983.

__________. News releases. San Jose, California, 1983.

Asimov, Isaac. *I, Robot.* Doubleday, New York, 1977.

__________. "Checking into Tomorrow's Hotel." *Signature,* November 1981, 54–55, 110–113.

Bridges, Les. "Brave New Work." *Success,* September 1983, 22–26.

Coates, Vary T. "The Potential Impacts of Robotics." *The Futurist,* February 1983, 28–32.

Cusack, Michael, "Industrial Robots." *Senior Scholastic,* February 4, 1983, 4–7.

Hapgood, Fred. "Inside a Robotics Lab: The Quest for Automatic Touch." *Technology Illustrated,* April 1983, 18–22.

__________. "Inside a Robotics Lab: Looking for Stereo Vision." *Technology Illustrated,* June 1983, 44–49.

Heath Company. "ET-18 Robot User's Guide." Benton Harbor, Michigan, 1982.

Mamis, Robert A. "The Pied Piper of Sunnyvale." *INC.,* March 1983, 57–66.

Parrett, Tom. "The Rise of the Robots." *Science Digest,* April 1983, 69–75, 107.

Solomon, Leslie. "Robotic Arms." *Computers & Electronics,* February 1983, 74–75.

GLOSSARY

Androbot, Inc.: The maker of Topo, B.O.B., F.R.E.D., and AndroMan. This company was formed by Nolan Bushnell, the founder of Atari, Inc.

Andromotion: Androbot, Inc.'s way of moving their personal robots from place to place. Andromotion is achieved by using three separate wheels to drive the probot and to turn it in any direction possible.

Applications Software: Programs designed for specific task, such as education or home protection.

BASIC: An acronym for Beginner's All Purpose Symbolic Instruction Code, the language used by most home computers and the one that controls such probots as Topo (TopoBASIC).

Bit: The smallest amount of information that a computer can hold. "Bit" is short for "binary digit," and a bit can be either on (represented by the digit 1) or off (represented by the digit 0). Eights bits, when used together, make a **byte**, which is a storage unit equivalent to one character of information.

Bug: A problem or mistake in a program.

Command: An instruction to the computer to perform a particular function.

Debug: To get rid of the mistakes or problems in a computer program.

Digit: Any number from zero to nine.

Documentation: The written instructions for a personal robot or a computer. Documentation usually includes books, quick reference cards, and other forms of written information so you can learn how to use your probot or computer.

Downloading: Sending a program by some electronic means.

Hardware: The actual mechanical and electronic parts of a robot or computer.

Heath, Inc.: Makers of the HERO 1 probot and many other electronic and electrical devices.

Hexadecimal: Base 16 numbers. We normally use the Base 10 system with digits from 0 to 9, but computers use hexadecimal numbers, ranging from 0 to 9 and A to F.

High-Level Language: A computer language, like BASIC that is easy to learn and use but is less efficient than the computer's own "low-level" language. This is because the computer must interpret the high-level words you are using and translate them into its own language.

Industrial Robot: Unlike a personal robot (or probot), this machine is usually an immobile, computer-controlled robot which does one repetitive task such as spray painting a car or welding two objects together.

Input: Information received by a machine, such as a computer or a robot.

Kilobyte or K: A total of 1,024 bytes of memory.

Language: The means of communicating. The computer's language, such as BASIC, is used to communicate with the mechanical robot.

Low-Level Language: A difficult-to-learn but extremely fast language, such as the HERO's "assembly language." Because it is in hexadecimal form, it's rather difficult for beginning and intermediate programmers to use, but it is very efficient.

Mobility: The power to move from one place to another.

Mode: A particular condition or state. For example, the "programming mode" would be a probot's condition of being ready to be programmed.

Output: Information sent from a machine, such as a computer or robot. If a probot uses its voice synthesizer, for example, to tell you something, it has given its output, or information, to you.

Probot: A personal robot.

Program: A set of instructions given to a computer.

Programmer: An individual who makes programs.

RAM: Random Access Memory, or the memory used to store a program or data in a computer or robot.

Reset: The process of re-initializing or starting something from the beginning.

Robot: A computer-controlled machine intended to serve a useful purpose. Robots usually have moveable parts and the ability to move around.

ROM: Read-Only Memory, used strictly by a computer for its own purposes. The language of the computer is permanently stored in ROM.

Sensor: A device used to detect something, such as heat, light, or sound.

Software: The programs a computer uses.

Tactile: Related to the sense of touch.

Ultrasonic: Sound which is beyond the limits of human hearing.

Uploading: Receiving a program by some electronic means.

Utility: A program designed to help a programmer.

Voice Synthesis: Computer-simulated speech.

INDEX

ABORT key (HERO 1), 34–35, 40
Advertising, probots used for, 100
Androbot, Inc., 9, 50, 67–68
  AndroMan robot by, 52, 88
  B.O.B. (Brains On Board) robot by, 68–72, 89
  F.R.E.D. (Friendly Robotic Educational Device) by, 52–55, 88–89
  Topo by, 55–65, 89
Androbot Control Language (ACL), 70
Androbus, 70
AndroFridge, 70
Androids, 50
AndroMan, 52, 88, 98
Andromotion, 56, 61
AndroSentry cartridge, 70
Androwagon, 70
Apple Computer, Inc., 9, 109
Apple II computers, 55, 94
  Topo controlled through, 57
Arms
  future of, 105
  of HERO 1, 31
    controlling, 37
  of industrial robots, 14, 16, 83–84, 94
  of probots, 5
  of RB5X robot, 77
  research and development of, 83–84
  uses of, 92–94
Artificial skin for robots, 84–86
Asimov, Isaac, 11, 112
Atari, 50
Atari 600XL computers, 9
Atari 2600 video computer system, 52
Auto mode (HERO 1), 41–43
Automobile industry, 19
Autonomy for robots, 68
Axes of rotation, 84

Babysitters
  B.O.B. as, 72
  HERO 1 as, 46
  probots used as, 96
BASIC (language)
  Tiny BASIC version of, 74
  TopoBASIC version of, 57–61, 64
Batteries for RB5X robot, 75
Blue collar workers, 19
B.O.B. (Brains On Board) robot, 68–71, 89
  uses of, 71–72
Bumpers (tactile sensors), 74–75, 84–86
Bushnell, Nolan, 49, 50, 70
Business, probot applications for, 100
Bytes, 42

Calculators, 104
Calibration of Topo, 61–64
Čapek, Karel, 10
Clocks in HERO 1, 31
Colossus of New York (film), 11
Commands
  for B.O.B., 70
  for HERO 1, 34–37, 41–42
  for Topo, 59
  voice-activated, 109
Commodore 64 computers, 9
Commodore VIC-20 computers, 9
Computers
  built into B.O.B., 68
  decline in price of, 104
  fears of, 12, 110
  negative effects of, 111
  robots compared with, 3
  on robots, uses of, 94
  (*See also* Home computers; Personal computers)
Convolvers, 86–88
Crime, probots used for, 111–112

Date, HERO's commands for, 37
Degrees of freedom of motion of robots, 84
Depth perception by robots, 86–88
Development of robots, 81–88
Displays on HERO 1, 31–34
Dog walking by probots, 96
Downloading of programs, 36

Education
  effects of robots on, 108
  probot applications for, 95
Electrical arms, 83–84
Employment
  created by robots, 22–25, 106–107
  industrial robots and, 17, 19–20
  robots and future of, 110
Entertainment
  AndroMan designed for, 52
  probot uses for, 98
Environments
  precautions regarding, 34
  Topo's movements through, 56
Executive mode (HERO 1), 34, 35
Expandability of robots, 70
Exploration of space and sea by robots, 112–113

Fiction, robots portrayed in, 10–11
Films, robots portrayed in, 11, 79, 89
Fire detection
  probots for, 98
  RB5X for, 81
Forbidden Planet (film), 11
F.R.E.D. (Friendly Robotic Educational Device), 52–55, 88–89, 98

Games
  AndroMan designed for, 52
  played by probots, 98
  video, 50, 52
Grippers (hands)
  of HERO 1, 31
  of industrial robots, 14, 16
  of probots, 5
  research and development of, 83
  (*See also* Arms)
Guard
  B.O.B. as, 70
  HERO 1 as, 46
  RB5X robot as, 79
  robot applications as, 93–94, 96–98

Hands on robots (*See* Grippers)
Head of HERO 1, 31
Heat sensors, 71
  used for sentry functions, 98
  uses of, 93
Heath Company, 9, 27
  HERO 1 robot by, 22, 28–47, 89
Heiserman, David, 74

HERO 1 robot, 22, 28-31, 89
operation of, 31-43
potential of, 46-47
speaking by, 43-46
HERO 2 robot, 47
Hewlett Packard, 106
Hexadecimal notation, 34, 41
History of robots and probots, 7-11
Hobbyists, 7
Home
future use of probots in, 108-109
use of probots in, 95-96
Home computers, 3, 6
history of, 7-10
peripherals for, 107
Topo controlled by, 55-56
(*See also* Computers; Personal computers)
Humans
anatomy of arms of, 84
fictionalized relationships between robots and, 10-11
future effects of robots on, 110
industrial robots and, 16-20
jobs for, created by robots, 22-25
robots and computers feared by, 12
Hydraulic arms, 83-84

I, Robot (short stories, Asimov), 11
Industrial Revolution, 19, 22
Industrial robots, 3, 13-25
advantages of, 17-19
arms on, 14, 16, 83-84, 94
artificial skin for, 86
disadvantages of, 19-20
expected drop in price of, 104
expected stages of development of, 20-21
Industries created by robotics, 106-107
Inflection of speech, 45
Initialization
of HERO 1, 35-36
of Topo, 59
INS 8073 microprocessor chips, 74
Intel 8086 microprocessor chips, 68
Intelligent robots, 72
used as industrial robots, 17-18
Interpreter language, 41, 43
Intruder alarm for RB5X robot, 79

Japan, industrial robots in, 16
Jobs
created by robots, 22-25, 106-107
robots and future of, 110
Jobs, Steve, 106
Joystick to move Topo, 57

Keypads, 34
Kits, HERO 1 as, 28
Kubrick, Stanley, 79

Labor
future effects of robots on, 110-111
(*See also* Employment)
Languages
Androbot Control Language (ACL), 70
interpreter, 41, 43
machine, 41-43
for RB5X, 77, 80, 81
Robot Control Language (RCL), 77, 81
taught by probots, 95
Tiny BASIC, 74
TopoBASIC, 57-61, 64
TopoLOGO, 64
Savvy, 77
Laziness as effect of robots, 112
Learn mode (HERO 1), 39-40
Light sensors, 93
on HERO 1, 30
LOGO (language)
for RB5X robot, 80
TopoLOGO version of, 64

Machine language, 41–43
Manual mode (HERO 1), 37
Manufacturing, industrial robots used for, 16–19
Markets, probots introduced into, 7–10
Media coverage for probots, 7
Memory
of B.O.B., 68
of HERO 1, 39, 42
during "sleep," 35
of probots, 5
of RB5X robot, 77
of robots, uses of, 94
Microprocessor chips
decline in price of, 81
INS 8073, 74
Intel 8086, 68
6800, 41, 42
Mining, robots used for, 113
Mobility
of probots, 5
RB5X package for, 81
uses of, 94
Motion detectors, 30, 93–94
Motions
degrees of freedom in, 84
detected by HERO 1, 30
of robot arms, 83
Music generated by RB5X robot, 79

Navigation detection by RB5X robot, 80

Party host
B.O.B. as, 71–72
probots used as, 96
Payloads carried by probots, 70
Pendant for HERO 1, 37, 39
Peripherals, 106–107
Personal computers, 3
F.R.E.D. controlled by, 55
manufactured by Heath, 28
Topo controlled by, 55–56
used to program RB5X robot, 74, 75
(*See also* Computers; Home computers)
Personal robots (*See* Probots)
Phonemes, 43–45
Pitch of HERO's voice, 46
Pneumatic arms, 83–84
Pocket calculators, 104
Pong (game), 49, 50
Power packs for RB5X robot, 77
Precautions for operation of HERO 1, 34
Prices
of microprocessor chips, 81
of robots, 88–90
expected drop in, 104
Probots (personal robots), 3
AndroMan, 52
B.O.B. (Brains On Board), 68–72, 89
comparisons of, 88–90
effects of, 107–111
features of, 5–6
F.R.E.D. (Friendly Robotic Educational Device), 52–55, 88–89, 98
future of, 103–107
HERO 1, 22, 28–47, 89
industrial robots and, 14–16
RB5X Intelligent Robot, 72–81, 89–90
stages of acceptance for, 7–10
Topo, 55–65, 89
uses for, 91–101
(*See also* Robots)
Productivity, 18
Program mode (HERO 1), 41–43
Programmability of probots, 5
Programming
of B.O.B., 70–72
of F.R.E.D., 55
of HERO 1, 34, 39–43
to speak or sing, 45–46
of industrial robots, 14, 18
of RB5X, 74, 75
of robots, future of, 109
of Topo, 57–61

Programs
for HERO 1
downloading, 36
uploading of, 36–37
for probots, 100–101
Psychological effects of robots, 110
Purbrick, John, 84, 86

R2-D2 robot (*Star Wars*), 28, 89
Radio communications for robots, 80
RB Robot Corporation, 72
RB5X Intelligent Robot of, 72–81
RB5X Intelligent Robot, 72–81, 89–90
add-ons, 77
features, 74–75
future of, 80–81
Repairs on robots
done by other robots, 111
jobs created by, 107
Repeat mode (HERO 1), 40
Research and development of robots, 81–88
RESET key (HERO 1), 34, 40, 43
RM arm for RB5X robot, 77
Robot Control Language (RCL), 77, 81
Robot Monster (film), 11
Robotics industry, 106
Robots
androids and, 50
AndroMan, 52
B.O.B. (Brains On Board), 68–72, 89
comparisons of, 88–90
computers compared with, 3
effects of, 107–111
employment created by, 22–25, 106–107
F.R.E.D. (Friendly Robotic Educational Device), 52–55, 88–89, 98
future of, 103–107
future uses of, 112–113
HERO 1, 22, 28–47, 89
industrial, 3, 13–25, 83–84, 86, 94, 104
negative effects of, 111–112
portrayed in fiction, 10–11
RB5X Intelligent Robot, 72–81, 89–90
research and development of, 81–88
Topo, 55–65, 89
(*See also* Probots)
Rotational axes, 84
Routines to move Topo, 57–61
RS-232 interfaces, 75
RTI (Reverse The Instruction) feature of HERO 1, 39–40
R.U.R. (play, Čapek), 10–11

Safety of robots, 110
Savvy (language), 77
Science fiction, robots in, 10–11
Sea exploration by robots, 113
Self-learning software, 74
Self-reproduction by robots, 21
Senses
of HERO 1, 30–31
touch, artificial skin for, 84–86
vision, 86–88
Sensors, 5
on B.O.B., 70, 71
on RB5X robot, 74–75
tactile, 84–86
uses of, 93–94
for sentry functions, 96, 98
Sentry
probots used as, 96–98
(*See also* Guard)
Shape recognition, 105
Silent Running (film), 11
6800 microprocessor chips, 41, 42
Skin, artificial, for robots, 84–86
Sleep switch (HERO 1), 35
Social effects of robots, 107–108
laziness as, 112
Software
in modules for RB5X robot, 75

Software (*cont.*)
for probots, 100–101
for RB5X robot, 74, 77–79
Sonar sensors, 74
Sound effects, 46
Sound sensors, 30, 93
Sound synthesis by RB5X robot, 77
Sounds
detected by HERO 1, 30
generated by HERO 1, 43–46
programming HERO 1 to respond to, 46
(*See also* Voice)
Space exploration by robots, 112–113
Speech
HERO's power of, 30, 43–46
of probots, 5
(*See also* Voice)
Star Wars (film), 11, 28, 89
Stereo vision for robots, 86–88
Subroutines to move Topo, 57–61

Tactile sensors (bumpers), 74–75, 84–86
Tandy Corporation, 9
Tape, programs saved on, 36
Teaching
effects of robots on, 108
probot applications for, 95
using B.O.B. for, 71
Teaching pendant for HERO 1, 37, 39
Telephones, 6, 92
Television, 6
Time, HERO's commands for, 37
Time circuits, 31
Tiny BASIC, 74
Topo probot, 55–57, 89
calibration of, 61–64
programming of, 57–61
uses for, 64–65
TopoBASIC, 57–61, 64
TopoFORTH, 64
TopoLOGO, 64
Topology, 56
Trailer for RB5X robot, 80
Travel, robots for simulation of, 108
2001: A Space Odyssey (film, Kubrick), 79

Ultrasonic range sensors, 30, 70, 71
uses of, 93
Unemployment, 19–20
future of robots and, 111
United States, industrial, 16, 18–19
Uploading of programs, 36–37
Utility mode (HERO 1), 35–37

Vacuum cleaner attachments, 80
VIC-20 computers (Commodore), 9
Video games, 50
robots and, 52
Vision by robots, 86–88
Voice
of B.O.B., 70
of F.R.E.D., 55
of HERO 1, 30, 43–46
of probots, 5
of RB5X, 77
robots commanded by, 109
of Topo, 56
uses of, 94
Voice recognition, 70
future of, 105, 109
by RB5X robot, 77, 80
Voice synthesis by RB5X robot, 77

Welding, 17
Work
future effects of robots on, 110–111
(*See also* Employment)
Wozniak, Steve, 106

Yamazaki Machinery Works, 17